Good Garden Bugs

First published in the United States of America in 2015 by
Quarry Books, a member of
Quarto Publishing Group USA Inc.
100 Cummings Center
Suite 406-L
Beverly, Massachusetts 01915-6101
Telephone: (978) 282-9590
Fax: (978) 283-2742
www.quarrybooks.com

10 9 8 7 6 5 4 3 2 1

ISBN: 978-1-59253-909-3

Digital edition published in 2015
eISBN: 978-1-62788-033-6

Library of Congress Cataloging-in-Publication Data
Gardiner, Mary M., author.
  Good garden bugs : everything you need to know about beneficial predatory insects / Mary M. Gardiner, Ph.D.
    pages cm
  Includes bibliographical references and index.
  ISBN 978-1-59253-909-3
  1. Beneficial insects. 2. Predatory insects. 3. Garden pests--Biological control. I. Title. II. Title: Everything you need to know about beneficial predatory insects.
  SF517.G37 2015
  635'.0496--dc23
                         2014049089

Design: Tony Seddon

Production Design: Barefoot Art Graphic Design

Cover Image: Laura Berman/Greenfusephotos.com

Illustrations: Michael Cooley

Photography: As noted

Printed in China

Good Garden Bugs

Everything You Need to Know About Beneficial Predatory Insects

Mary M. Gardiner, Ph.D.

Quarry Books
100 Cummings Center, Suite 406L
Beverly, MA 01915

quarrybooks.com • wholehomenews.com

Contents

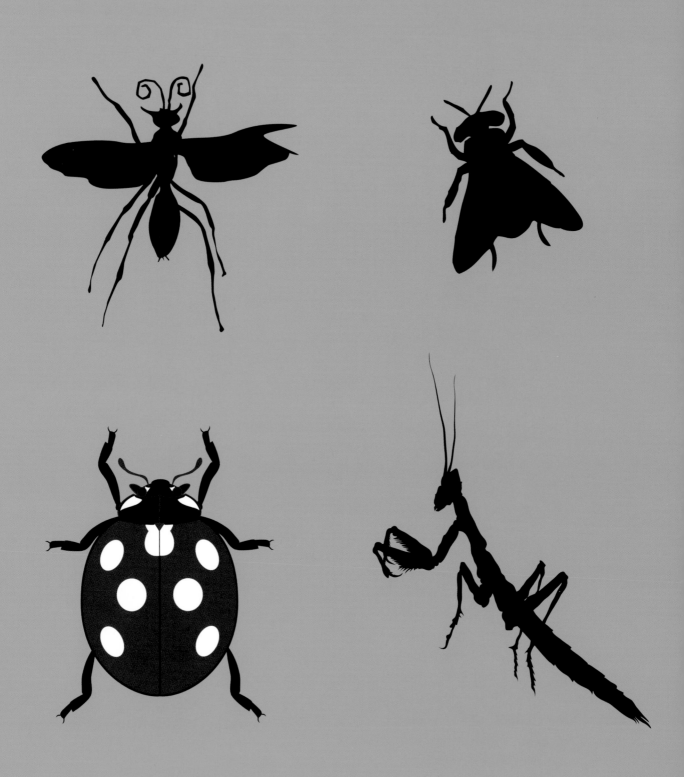

Introduction: Knowing the Good Garden Bugs

Many of us are all too familiar with garden pests and the damage they inflict on our favorite plants. We see evidence of their destruction during daily walks through the garden, whether it's rose leaves that have turned rusty from spider mites, the curled leaves of a spirea bush that is covered with aphids, or withered squash plants that have contracted bacterial wilt, a disease transmitted by voracious cucumber beetles. Of course, these and other pests that damage plants and spoil vegetables are the bane of many gardeners. And, when encountering the damaging effects of insects in the garden on a regular basis, it's easy to lose sight of how helpful some insects can be. But believe it or not, many insects actually provide beneficial services to gardeners.

In this book, I will introduce you to an important group of good garden bugs – those that provide natural pest control in home landscapes. Throughout the book, I'll refer to these good bugs as natural enemies. From lady beetles to spiders, and lacewings to wasps, these insects include some of the creepiest, scariest, and, in my opinion, most misunderstood (and most likely to be stepped on) backyard creatures. These helpful arthropods are natural enemies of the pests they attack. So before you put your foot down, consider harnessing these natural enemies for the good of your garden, reducing your need to rely on harmful chemicals in the process. On the pages of this book, you'll find everything you need to know to identify and attract natural enemies to your garden. And, as you learn more about them, I hope that you will find their rich diversity, unique behaviors, and distinctive beauty as interesting as I do.

How to Use This Book

In my experience, many gardeners have a natural curiosity about the insects they find among their plants. Not only do they have the desire to learn how to identify specific bugs, but they are also intrigued by what these insects are doing in the garden on a day-to-day basis. This is important because the support natural enemies provide in the garden can vary according to where they may be in their lifecycle. Some insect species, for example, consume the most garden pests while in the larval stage, while others are most helpful as adults. This comprehensive guide includes information about how to identify natural predators in your garden in addition to important details about the predator's life history attributes and behavioral traits. And I couldn't resist including a few odd creatures that are more scarce, but fascinating. These "odd bugs" are a reminder of the incredible diversity within this group of arthropods.

Chapter Organization

Before diving into the fauna of backyard natural enemies, here is a brief overview of how the natural enemy identification chapters are organized. If you remember learning the levels of the animal classification chart in sixth-grade science, the order of the chapters will make perfect sense. If not, think of this as a refresher course!

All arthropods are in the phylum Arthropoda. Within Arthropoda, I will discuss natural enemies that belong to three classes: Insecta (all insects), Arachnida (all arachnids), and Chilopoda (centipedes). These groups are further broken down into orders, families, genera, and species of arthropods. In chapters 4 through 9, you'll find information on garden bugs that fall into the class Insecta, with each

A Word about the Photos

chapter focusing on one insect order. Chapters 10 and 11 introduce the class Arachnida, with chapter 10 discussing spiders in the order Araneae, while chapter 11 includes information about several orders of interesting arachnids such as mites (Acari), harvestmen (Opiliones), and scorpions (Scorpiones). Finally, chapter 12 focuses on the Chilopoda, or centipedes.

Each chapter describes important families, genera, and species within each order. Family names can be distinguished from other levels of biological classification because they always end in "idae," as in the spider families Salticidae (jumping spiders), Linyphiidae (sheetweb spiders), and Oxyopidae (lynx spiders). Within an arthropod family, the genus and species names for specific arthropods are italicized, and the genus name begins with a capital letter while the species name begins with a lowercase letter. For example, the bold jumping spider *(Phidippus audax)* is a common species within the spider family Salticidae. In addition to the scientific names of the insects, terminology specific to the morphology and behavior of arthropods are defined within the text for your convenience. I also indicate the types of arthropods attacked by each group. Many of these species are generalists, meaning they feed on a number of different prey. When a specific family, genus, or species of prey is listed, I provide the scientific name to make looking up images and more information easier.

It can be very difficult to recognize a tiny insect in the garden from a large photo. Make sure to note the size of each arthropod, which is given millimeters. For larger creatures, a conversion to inches is also provided.

What good is an identification guide without pictures? The majority of the arthropod images that appear on these pages are from the website bugguide.net, which is hosted by the Department of Entomology at Iowa State University. This website is an amazing resource where anyone can post images of insects they've discovered in their gardens. Experts monitor the website and provide identifications and information about the submitted arthropods. The photographers featured in this book range from professionals to homeowners who found something in their backyard worth snapping and submitting. It was not possible to include a photo of every insect mentioned in this book, but I encourage you to search bugguide.net for additional images. And, feel free to submit your own photos to identify natural enemies and learn more about them.

Photo: Barbara Thurlow

△ *Good garden bugs like this assassin bug provide a beneficial service to gardeners by consuming pest insects. They also contribute to the amazing biodiversity we can find in our own backyards.*

Arthropod Classification Chart

This animal classification chart includes common names for each insect and specifies the chapter in which details about the insects of each order can be found.

Kingdom: Animalia (animals)
   Phylum: Arthropoda (arthropods)
     Class: Insecta (insects)
       Order: Mantodea (chapter 4: mantids, page 39)
         Family: Mantidae (Mantids)
       Order: Hemiptera (chapter 5: true bugs, page 45)
         Family: Anthocoridae (minute pirate bugs, page 46)
         Family: Belostomatidae (giant water bugs, page 56)
         Family: Corixidae (water boatman, page 57)
         Family: Geocoridae (big-eyed bugs, page 47)
         Family: Gerridae (water striders, page 57)
         Family: Nabidae (damsel bugs, page 47)
         Family: Naucoridae (creeping water bugs, page 58)
         Family: Nepidae (water scorpions, page 58)
         Family: Notonectidae (backswimmers, page 59)
         Family: Pentatomidae (predatory stink bugs, page 48)
         Family: Reduviidae (thread-legged, ambush, and assassin bugs, page 50)
       Order: Neuroptera (chapter 6: lacewings, owlflies, and antlions, page 60)
         Family: Chrysopidae (green lacewings, page 62)
         Family: Hemerobiidae (brown lacewings, page 64)
         Family: Coniopterygidae (dustywings, page 64)
         Family: Mantispidae (mantidflies, page 65)
         Family: Ascalaphidae (owlflies, page 61)
         Family: Myrmeleontidae (antlions, page 67)
       Order: Coleoptera (chapter 7: beetles, page 68)
         Family: Cantharidae (soldier beetles, page 69)
         Family: Carabidae (ground and tiger beetles, page 71)
         Family: Coccinellidae (lady beetles, page 79)
         Family: Dytiscidae (predaceous diving beetles, page 88)
         Family: Gyrinidae (whirligig beetles, page 89)
         Family: Lampyridae (fireflies, page 84)
         Family: Melyridae (soft-winged flower beetles, page 85)
         Family: Staphylinidae (rove beetles, page 86)
       Order: Diptera (chapter 8: flies, page 90)
         Family: Asilidae (robber flies, page 92)
         Family: Cecidomyiidae (predatory midges, page 94)
         Family: Dolichopodidae (long-legged flies, page 95)
         Family: Empididae (dance flies, page 96)
         Family: Rhagionidae (snipe flies, page 97)
         Family: Sciomyzidae (snail-killing or marsh flies, page 98)
         Family: Syrphidae (hover or flower flies, page 98)
         Family: Tachinidae (parasitoid flies, page 100)
       Order: Hymenoptera (chapter 9: wasps and ants, page 102)
         Family: Aphelinidae (aphelinid wasps, page 108)
         Family: Bethylidae (bethylid wasps, page 110)
         Family: Braconidae (braconid wasps, page 104)
         Family: Chalcididae (chalcid wasps, page 105)
         Family: Crabronidae (sand wasps, square-headed wasps, aphid wasps, and beewolves, page 110)
         Family: Encyrtidae (encyrtid wasps, page 108)
         Family: Eulophidae (eulophid wasps, page 108)
         Family: Eurytomidae (eurytomid wasps, page 108)
         Family: Formicidae (ants, page 125)
         Family: Ichneumonidae (ichneumon wasps, page 105)
         Family: Mutillidae (velvet ants, page 114)
         Family: Mymaridae (fairyflies, page 108)
         Family: Platygastridae (platygastrid wasps, page 108)
         Family: Pompilidae (spider wasps, page 115)
         Family: Pteromalidae (pteromalid wasps, page 108)
         Family: Scoliidea (flower wasps, page 112)
         Family: Sphecidae (thread-waisted wasps, page 116)
         Family: Trichogrammatidae (trichogrammid wasps, page 108)
         Family: Vespidae (yellow jackets, hornets, paper wasps, mason wasps, and potter wasps, page 118)

1

The Basics:
Natural Enemy
Lifecycles and
Eating Habits

Natural enemy arthropods vary widely in the prey they seek, the environment in which they live, and their lifecycles. Although these specific details are presented in the individual chapters, I wanted to draw your attention to two key distinctions in how natural enemies develop and how they feed.

Knowing the lifecycle of important natural enemies is helpful as you begin to identify these insects in your garden. When armed with this knowledge, you'll have a better understanding of how natural enemies impact pests in the garden. The natural enemy arthropods can be divided into two different groups: those that develop via gradual metamorphosis and those that undergo complete metamorphosis.

Gradual Metamorphosis

All orders in the classes Arachnida and Chilopoda, as well as the two orders within the class Insecta included in this book — Mantodea (mantids) and Hemiptera (true bugs) — develop via gradual metamorphosis. This means that they hatch from an egg into an immature stage typically called a nymph (spider hatchlings are also referred to as spiderlings and juveniles). These arthropods undergo several molts to reach the reproductive adult stage, but they do not have a pupal stage, so will you will never find a cocoon of a spider or mantid in the garden. The nymphs will become larger following each molt, and if the arthropod is winged as an adult, its wings will gradually form. These partially formed wings are called wing buds and are not functional until the adult stage. Typically, the nymph or immature stages of these groups look similar to the adult, although with some species, picking out the family resemblance is much easier when immatures are close to reaching the adult stage. Both the immature stages and adults of these arthropods will typically feed on garden pests. Often, there is also considerable overlap in the types of prey immatures and adults will consume. For example, nymph and adult minute pirate bugs (Anthocoridae) feed on many common garden pests, including aphids (Aphididae), spider mites (Tetranychidae), and insect eggs. In most cases then, you can find both nymphs and adults foraging within the same habitat, such as on the soil or crawling on plant leaves, and you often will find both nymphs and adults present at the same time in the garden.

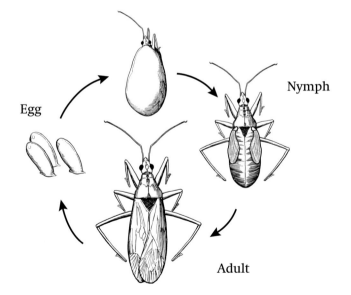

Egg

Nymph

Adult

△ *Natural enemies that develop via gradual metamorphosis do not have a pupal stage. These arthropods hatch from an egg into an immature stage called a nymph. The nymph will undergo several molts, growing larger each time, and if they will be winged as an adult, their wings will form slowly during their development. After its final molt, a nymph will become an adult able to reproduce, and if winged, its wings will be fully formed. When insects develop via gradual metamorphosis, both the nymph and adult stage typically feed on prey within the garden.*

Complete Metamorphosis

Four orders of natural enemies discussed in this book develop via complete metamorphosis: Neuroptera (lacewings, owlflies, and antlions), Coleoptera (beetles), Diptera (flies), and Hymenoptera (wasps and ants). Here, insects emerge from the egg as a larva, which develops through several molts, increasing in size each time. The larvae of insects that undergo complete metamorphosis often look very different from the adult stage. They are usually somewhat worm- or caterpillar-like – they do not have wings (and developing wing buds are not present) and may or may not have legs. Eventually, insects with complete metamorphosis will construct a pupa (or cocoon), and inside this protected case, they will undergo a major transformation. When they emerge, they look much different from their larval stage; most have wings and have a different general body form. They are also able to reproduce.

Arthropods that develop through complete metamorphosis are more likely to vary in what they eat among life stages than those that develop by gradual metamorphosis. The immature, larval stages of these natural enemies are typically predacious. A larva may find its arthropod prey on its own or be provided with food by adults. For example, the larvae of lady beetles (Coccinellidae) actively seek out aphids, whereas an adult female thread-waisted wasp (Sphecidae) will collect arthropods for its larvae to eat. Adults may or may not feed on prey themselves. Also, the larvae and adults are not always found in the same place within the garden and may not hunt in the same way. For example, the larvae of tiger beetles (Carabidae) construct tunnel-shaped burrows into the soil and wait inside to catch unsuspecting insect prey, whereas the adults run along the soil surface searching for prey.

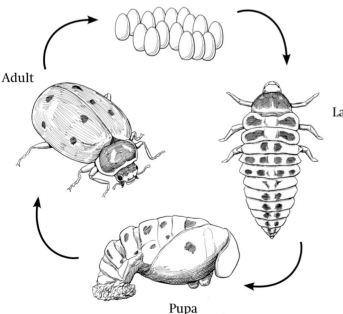

Egg

Larva

Adult

Pupa

◁ *Insects that undergo complete metamorphosis will have egg, larval, pupal, and adult stages. A larva will emerge from its egg and, over the course of this life stage, undergo several molts, gaining size each time. Often larvae are slender bodied. They can appear wormlike without legs, or they may have three pairs of legs present. Eventually, the larva will be ready to construct a pupa and enter this resting stage where major transformation occurs. When ready, the insect will emerge from the pupa as an adult. Typically, insects that develop via complete metamorphosis look very different during the adult and larval stages. Adults may also be found in a different part of the garden than the larva, such as above versus below ground, and may or may not feed on insect prey.*

Eating Habits: Predators and Parasitoids

A second major distinction among natural enemies is whether they feed as a predator or parasitoid. Each of the arthropod chapters in this book includes predatory species. Living as a parasitoid is most common among wasps (Hymenoptera) and to a lesser extent flies (Diptera) and beetles (Coleoptera); it is rare to nonexistent within other groups of arthropods.

Predators

Predators are arthropods that actively hunt for their own prey. Most predatory arthropods are generalists, meaning that they will find many different types of insects acceptable as food and will also eat other types of food, such as pollen or nectar from flowers.

Some generalist predators, such as lady beetles (Coccinellidae), consume mainly pest insects, including aphids, scales, and mealybugs. Others, such as many spiders (Araneae), will consume both pests and other beneficial insects. Since pests are often common in home gardens, it is likely that these unwanted insects will make up much but maybe not all of the diet of many of the predators you encounter.

Predators vary widely in their hunting strategy. Many spend their days walking and flying around the garden and other surrounding habitats in search of food. These are called active hunters and examples include wolf spiders (Lycosidae) and minute pirate bugs (Anthocoridae). Active hunters with large eyes such as jumping spiders (Salticidae) may use visual cues to locate prey, whereas others may track prey using olfactory or tactile cues. For example, aphid predatory midge larva (Cecidomyiidae: *Aphidoletes aphidimyza*) cannot see, thus as they crawl along on leaves, they rely on feeling the leg of an aphid to know they have found their next meal. Then the larva will pierce the leg joint, paralyze the aphid, and ingest the aphid's liquid contents.

Other predators, such as crab spiders (Thomisidae) and most mantids (Mantidae), wait for their next meal to find them. These sit-and-wait predators are able to remain motionless until prey comes within their grasp; then they move quickly to capture it. Often, these predators are well camouflaged to their hunting grounds. Web-building spiders, such as orb weavers (Araneidae) and sheetweb spiders (Linyphiidae), take the sit-and-wait strategy a step further by creating a trap to catch passing prey.

Parasitoids

Parasitoids live their lives quite differently from predators. This group of insects is similar to parasites in that they live internally during part of their lifecycle. Yet unlike a parasite where the host animal survives, as they develop, parasitoids will eventually kill their host.

The life of a female parasitoid centers on locating a host arthropod for her offspring to feed upon. She will lay eggs within, on, or near this host using an ovipositor, which is a tubular, egg-laying organ. For some parasitoid wasps, the ovipositor is clearly visible and can be longer than the body of the insect! Ovipositor length varies most widely among wasps, and the length can be an indicator of how difficult it is for the parasitoid to get to its host. Those such as the giant ichneumons (Ichneumonidae: *Megarhyssa*), which attack large prey or prey hidden inside stems, seeds, or even under bark, have longer ovipositors. It's a good thing that parasitoid wasps cannot sting humans with their ovipositor and only are a threat to their targeted host insects!

Most parasitoids are considered "specialists" in that they will attack only one or a few species of host arthropods. Some females search for a specific species of arthropod, while others will accept several different hosts, but usually in the same arthropod family or order. A small parasitoid flying around in a garden looking for a specific host might seem akin to looking for a needle in a haystack, but female parasitoids use a lot of cues to narrow in and locate their host. When determining where fresh eggs may be located, parasitoids are able to follow odors, such as those from the droppings of caterpillars chewing on a corn plant or even the mating pheromones of adult hosts. For concealed hosts, such as those feeding inside rolled leaves or under bark, vibrational cues are also very important.

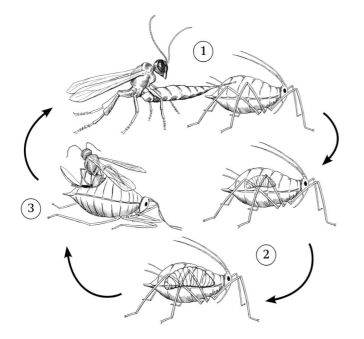

1. *A female parasitoid seeks a host as a food source for her offspring. Many species attack aphids (Aphididae). This female wasp is inserting one egg inside the pest with her ovipositor.*

2. *The egg of the wasp hatches within the aphid and the larva consumes its internal contents.*

3. *The wasp larva will pupate within the dead aphid, which darkens and hardens into an aphid mummy. Often, aphid mummies are the only visible evidence of aphid parasitism you will see in the garden. When ready to emerge as an adult, the parasitoid chews a round exit hole through the dead aphid and flies off in search of a mate.*

Once the host is located, the female parasitoid uses her ovipositor to lay one or more eggs. Some wasp parasitoids will pierce the host and deposit one or more eggs directly within it. Other parasitoids lay their eggs on the "skin" or exoskeleton of the host insect. Then the larvae will hatch and burrow into the host before feeding. Females may also lay eggs on plant material, where they will be ingested by the host. Some of these eggs are not destroyed during feeding and will hatch within the gut of the host. In other species, females lay eggs throughout the environment, and the larvae are left to find their own suitable host. One host can support a single or many parasitoids. In some species, each egg is laid within an individual host, and only one larva per host will survive to reach adulthood. In other cases, many eggs can be deposited into a single host. In some instances, parasitoids can produce one egg that divides, resulting in the production of several larvae in a process called polyembryony. Interestingly, some parasitoids can even select the sex of their offspring when depositing eggs. Female wasp parasitoids can deposit unfertilized male eggs in smaller, less-ideal hosts and provide fertilized female eggs with the most choice food supply for development.

If a parasitoid deposits its eggs inside of the host, its larvae will develop as an endoparasitoid, meaning inside the host. These larvae are typically light in color and legless. If eggs are deposited outside of the host, the larvae either burrow into the host themselves or feed on it externally as an ectoparasitoid. In some species of parasitoids, the host is killed or paralyzed with oviposition, and the larvae consume it in the location where it was attacked. In other cases, the host does not die immediately and may continue to feed, molt, and perform other normal functions as the larvae of the parasitoid consume its internal organs and tissues. The host will eventually die when the larvae of the parasitoid reach the pupal stage. Parasitoid larvae may pupate within the body of the dead host, or they will emerge and construct their pupa on or near the host.

When the adults emerge from the pupal stage, they seek food and mates. They will feed on pollen and nectar and aphid honey dew, which is a sugar-rich liquid that aphids secrete as they feed on plants. Some female wasps also gain nutrition through "host feeding" in which a female will sting a host with her ovipositor and then feed on the blood and tissue in the wound. She typically does not lay an egg inside hosts where host feeding occurs.

These differences in development and feeding only begin to illustrate the diversity of natural enemies. The chapters that follow dig into the "secret lives" of these good garden bugs to reveal more about how they forage, court mates, care for young, build shelters and nests, and much more.

2

Biological Control in Home Gardens

With the right balance of plant-feeding herbivores and natural enemies, you can create a sustainable garden at home. Gardeners can achieve this balance in part by applying biological control tactics, which range from releasing natural enemies to conserving those already present within your landscape. In this chapter, I describe how importation, augmentation, and conservation biological control can be useful to gardeners who are trying to manage home landscape pests.

Importation and augmentation biological control both focus on releasing natural enemies into the landscape. Importation involves the release of nonnative natural enemies in order to reduce populations of invasive pests accidently introduced into the United States. As gardeners, we are not going to perform these importation biological control releases ourselves, but we benefit from successful programs that reduce damage to garden and landscape plants. We can perform our own augmentation biological control releases, where naturally occurring predators or parasitoids can be purchased and introduced into the garden to combat pest infestations.

When thinking about applying biological control practices in a home landscape, however, conservation biological control is the place to start. Conservation biological control involves building up existing natural enemy communities by creating a suitable habitat that provides all the resources they need to thrive. Application of these conservation practices can mitigate the need to take further action, whether that is the release of biological control agents or the use of chemicals to control insect pests.

Also consider other nonchemical control methods that can be used with biological control to manage pests. This can be as simple as removing the pests by hand. For small gardens, dropping pests into a bucket of soapy water works well for immobile or slow-moving pests such as pest eggs and many larvae. For other very mobile and damaging pests, such as striped and spotted cucumber beetles (Chrysomelidae: *Acalymma vittatum* and *Diabrotica undecimpunctata*), try covering the plants with row covers to prevent the beetles from accessing their food source. These covers can be placed over the crop and may be used with or without hoops (when used without hoops, they are called floating row covers). For crops that require bee pollination, however, you will need to remove the cover when the plant is flowering. This gives the plant a pest-free window prior to flowering, with biological control coming into play after the row cover is removed.

Photo: Jim Jasinski

△ *Biological control can be combined with other management practices such as the use of row covers to create a sustainable home garden. This row cover is attached to a raised bed.*

Importation Biological Control

Importation biological control targets species that have been accidently introduced into the United States. These pests are referred to as invasive, meaning a non-native species that causes significant harm to our environment. Examples of such pests include the emerald ash borer (Buprestidae: *Agrilus planipennis*), which has decimated our native ash trees, and the red imported fire ant (Formicidae: *Solenopsis invicta*), which have displaced native ants and, when encountered in home landscapes, will sting people and pets. These invasive pests arrive in many ways, including on infested nursery stock or in the surrounding soil, on imported fruits and vegetables, or, as is the case with wood-boring insects, within pallets of untreated lumber or furniture. And, like many invasive plants, some have been intentionally released. Aquatic invaders have also been known to hitch a ride within the ballast water of ships. The vast majority of these invaders are detected when the contaminated shipments arrive and are inspected in a United States' port of entry. Unfortunately, however, the small number of invasive species that make it past this process can cause considerable damage to the balance within existing ecosystems.

When an invasive species establishes itself in the United States, government agencies and university researchers attempt to eradicate it when populations are small and this type of intensive effort can be successful. But if eradication fails and the species becomes abundant over a large area, goals shift to managing the population at as low of a level as possible. Importation biological control is one method that can be used to reduce the abundance of an invader. Typically, invasive organisms are not nearly as abundant and damaging in their native range due to suppression by natural enemies. When they are released into a new area, however, where their natural enemies are not around to keep them under control, they can cause devastating damage.

In importation biological control, researchers travel to the home range of the invader and search for natural enemies of the invasive pest. Any natural enemies found during these explorations are brought to the United States, and colonies are established within a quarantine facility. In this facility, years of research will be conducted to evaluate the collected natural enemies for possible release. This is an intensive process because it is critical to determine if the natural enemy will attack other native species that may or may not be closely related to the invasive pest.

Only natural enemies found to be host specific—meaning that they only consume the target invasive pest—will be considered for release. For this reason, the natural enemies of arthropod pests considered for release as part of an importation biological control program are nearly always parasitoids. While predators may be effective as well, they are far more likely to be generalists that will attack a broad range of arthropods, including native species. If, after intensive testing, a biological control agent is approved by the U.S. Department of Agriculture for release, the natural enemy will be reared in quarantine to obtain enough individuals to release several small populations. Follow-up studies are then done to determine if these populations establish and are able to effectively control the target pest.

The release of biological control agents was not always evaluated as carefully and extensively as it is today. These protocols were developed to prevent the introduction of non-native natural enemies that consume native species in addition to the target invasive pest, as has happened following some previous releases. These negative effects, called nontarget impacts, can also include competition between native species and released natural enemies for food or other resources. Due to stricter regulations, the number of importation biological control releases has decreased over time, but the proportion of species released that have a significant impact on their target invasive pest has remained relatively consistent.

△ *Graduate student Dana Roberts inspects logs infected with Asian longhorn beetles (Cerambycidae:* Anoplophora glabripennis) *within a quarantine facility at Penn State University. Before any parasitoid wasps can be released to suppress this damaging beetle, they must be evaluated in quarantine to determine their effectiveness and any possible negative nontarget impacts.*

Augmentation Biological Control

This is the type of biological control that is most familiar to gardeners. Augmentation biological control involves the periodic release of natural enemies that already occur naturally in your region by individuals, including gardeners and farmers. The idea behind this pest management strategy is that in some cases, existing natural enemy communities are not sufficient to control garden pests; perhaps they have been negatively affected by practices such as tillage or pesticide application, or simply have not yet colonized a garden in sufficient number. With augmentation, the goal is to supplement the existing population with enough natural enemies to enhance pest control.

There are two types of augmentative releases: inundation or inoculation. Gardeners may conduct inundation releases when they have a significant pest problem and are releasing a large number of natural enemies to suppress the pest outbreak. This approach garners fast results by quickly overwhelming the pest population, but this large number of natural enemies is likely to die off or leave the area after exhausting all sources of food. So inundation can be effective when pest populations are very high, but if the pest population continues to grow or reestablishes later, additional releases may be necessary.

In an inoculation release, a small number of natural enemies are released with the idea that these individuals will establish within the garden and reproduce. Some control will be provided by the released individuals, but the real benefits will come from subsequent generations. To establish a natural enemy population that will survive and flourish in a home garden, it is critical to consider how you manage your garden habitat to support these arthropods. See Conservation Biological Control on page 23 for tips on how to make your garden more inviting to beneficial insects.

When examining your options for augmentative biological control releases, first identify your target pest. If you are not sure what type of insect you are dealing with, take a sample of the damaged plant material and the pest itself to a nearby university extension office or reputable local garden store for identification. Some of the most common insects controlled through augmentative releases include aphids (Aphididae), scales (Coccoidea), mealybugs (Pseudococcidae), spider mites (Tetranychidae), thrips (Thripidae), and caterpillars (Lepidoptera). Next, determine if purchasing and releasing natural enemies is worth the investment. To do this, it is important to monitor your plants frequently. If you notice that an insect pest is beginning to multiply, look for evidence of natural enemies at work. Do you see predators foraging on the infested plant? For pests that do not move far, such as aphids (Aphididae) or caterpillars (Lepidoptera), do you see fewer pests present over a few-day period? Or perhaps there is evidence of parasitism, such as some aphids within a patch turning to brown or black aphid mummies? If so, releasing natural enemies may not be necessary.

If you do not see evidence of naturally occurring biological control, you can consider augmenting the population by purchasing natural enemies. Typically, gardeners purchase natural enemies from a garden center, home improvement store, or from a biological control supply company found online. A wide range of species are commercially available, including parasitoids and predators.

Parasitoids for Augmentation Releases

If releasing a parasitoid, it is critical to purchase the correct species that will attack your pest. Most parasitoids are very selective about the hosts they will attack. A reputable biological control company will clearly state the parasitoid's host range, lifecycle information, recommended release rates, and instructions for the release, with tips on how to evaluate the effectiveness of the natural enemies.

Before purchasing parasitoids, check the extended weather forecast in your area to make sure rain or extreme heat is not expected. Most companies advise against holding onto your shipments for more than a couple of days, so you want to be sure conditions will be favorable for release before purchasing the insects. Parasitoids are often shipped in the pupal stage, but some may have already emerged as adults. For best results, make sure to read and follow all the provided release instructions carefully. Your pupae may arrive on cards that are specifically designed to be attached to plants, or they may be sent in a bottle mixed with a carrier such as woodchips, seed husks, or vermiculite.

Photo: MaLisa Spring

△ *Wasp pupae and emerging adults are often sent within a bottle containing a carrier. The bottle is designed to allow you to shake the pupae onto or near infested plants. Because the pupae are small and can be very difficult to see, the carrier will allow you to see where you have dispersed them.*

TIP: *Events during shipment can affect the emergence of wasps; therefore, it is advisable to hold back a few individuals in a container with a fine mesh lid. Place this container out of direct sunlight and monitor it to determine if the wasps were alive upon receipt and emerged within the timeframe stated by the company.*

Predators for Augmentation Releases

Many generalist predators, including the spined soldier bug (Pentatomidae: *Podisus maculiventris*), minute pirate bugs (Anthocoridae: *Orius*), green lacewings (Chrysopidae), lady beetles (Coccinellidae), hover flies (Syrphidae), and predatory mites (Acari), are available commercially. Predators may be shipped as eggs, nymphs or larvae, pupae, or adults. Depending on the natural enemy, one or more life stages will be available. When deciding what life stage to order, consider whether both adults and immatures (nymphs or larvae) feed on prey. If only the immature stage is a meat eater and you release adults, you will have a delay before immatures from the next generation begin feeding. This isn't necessarily a bad thing if you are trying to build up a population, but it is

something to consider if the pest infestation is significant and plant damage or loss is already occurring. Also, adults and immatures might differ in their ability to disperse. Immature arthropods lack functional wings, so they are more likely to stay and feed where you release them, whereas adults may or may not be winged.

Predators may be sold as eggs on cards or as immatures, pupae, or adults in containers. Sometimes adults might be shipped with a sugar-water solution or other food source, or, like parasitoid pupae, they may be sent in carrier such as vermiculite, seed husks, or woodchips. Make sure to read and follow the directions for release carefully to know how and when to introduce your predators to the garden.

Photo: MaLisa Spring

△ *Natural enemies can face harsh conditions on their way to the garden, such as being left on a hot porch while you are at work. To ensure they have arrived alive and hungry for garden pests, inspect your shipment carefully. If you receive an active stage, shake a small amount of the carrier onto a white sheet of paper to determine if the predators are alive and in good condition. If you order eggs or pupae, allow a small sample to emerge within a container with a fine mesh vent, kept out of direct sunlight to ensure their quality.*

Avoid Releasing Convergent Lady Beetles

The convergent lady beetle (*Hippodamia convergens*) is the most common natural enemy sold in home improvement stores and garden centers. Unlike nearly all other natural enemies, these beetles are collected in the wild, and although the species is present throughout the United States, it is collected nearly exclusively from the foothills of the Sierra Nevada Mountains in California. This lady beetle aggregates in very large numbers to spend the winter in these foothills, and in the spring, they disperse to seek prey within crop fields, natural habitats, and home landscapes. Collectors visit these aggregations and are able to shovel thousands of individuals into buckets to be sold to biological control companies. The convergent lady beetle is very abundant in California, and there is currently no evidence that these collections are damaging populations within the region. However, in my opinion, there are ecological and practical reasons to avoid purchasing and releasing convergent lady beetles. First, wild-collected populations are not guaranteed to be free of diseases or parasitoids. Releasing these lady beetles could mean releasing their afflictions as well, which could impact established convergent lady beetle populations in your region that may or may not have been exposed previously. Second, from a practical standpoint, gardeners are likely to get little pest control for their investment. These lady beetles were collected from an over-wintering site, thus, when they are released into places with warm spring and summer temperatures, their instinct is to fly away in search of food. So, this investment might help others, but these beetles will likely leave your garden without consuming a significant number of pests.

Although I do not know of any sources of convergent lady beetle that are not wild collected, other lady beetles are raised in colonies by biological control companies or their suppliers. These species, such as the mealybug destroyer (*Cryptolaemus montrouzieri*) and mite destroyer (*Stethorus punctillum*), should therefore be free of parasitoids and diseases. A reputable supplier will provide information on how your natural enemies were collected or reared if you have concerns prior to purchasing.

▷ *Convergent lady beetles are shipped as adults, typically in a small mesh or cotton bag. By far, they are also the most common natural enemy you will find in stores.*

Photo: MaLisa Spring

Conservation Biological Control: Designing a Natural Enemy–Friendly Landscape

Gardening practices such as planting, mulching, weeding, pesticide application, and fruit and vegetable harvesting can disturb natural enemy communities and affect the amount of biological control they can provide in a home landscape. Conservation biological control focuses on minimizing the impacts of these practices by adding the resources natural enemies need to thrive locally.

Establishing Insectary Plantings

Insectary plantings are a great way to support a diverse and abundant natural enemy population within a home landscape. Insectary plantings are groupings of annual and/or perennial plants that sustain and enhance populations of beneficial arthropods — both natural enemies and pollinators — by providing resources such as pollen and nectar, alternative prey, winter cover, and habitat for nesting and retreats.

Nearly all gardeners know that bees visit flowers to feed on and collect protein-rich pollen and sugar-rich nectar, but few recognize the importance of these resources for predators and parasitoids. Many groups of natural enemies feed on pollen and nectar, including many beetles (Coleoptera), true bugs (Hemiptera), lacewings (Neuroptera), predatory wasps (Hymenoptera), parasitoids (Diptera and Hymenoptera), and even spiders (Araneae). Access to pollen and nectar increases their longevity and egg production, improving the ability of a natural enemy community to provide biological control. For some, access to these resources is essential—parasitoid wasps, for example, cannot survive and reproduce without access to flowering plants.

Photo: Julie A. Peterson

△ *Spiders are voracious predators, but they actually eat a more diversified diet than most people realize. Active hunting spiders can visit flowers and consume pollen directly. Those that build webs may do this as well, or consume pollen grains that land within their webs, as shown in the web of this sheetweb spider (Linyphiidae).*

Not only do natural enemies benefit from the foods plants naturally produce, but also from the herbivores found feeding on them. Many times, the herbivores that feed on flowering garden plants are not the same pests that can build up in large numbers within fruit and vegetable gardens. When gardeners provide a diversity of flowering plants, natural enemies are able to reproduce and multiply within a home landscape by feeding on these arthropods and are then present to begin feeding on pests that may arrive.

When designing an insectary habitat in a home landscape, the goal is to provide both the food and shelter resources natural enemies need to survive year-round. Natural enemies require floral resources throughout the growing season, so selecting several different flowering plants to create a consistent supply of blooms across the season is key. It's especially important to look for early spring and late fall blooming species because it is at these times of the year that resources tend to dwindle in home landscapes. Also, since many natural enemies do not have the specialized mouthparts that allow some bees and butterflies to gather nectar and pollen from deep flowers, make sure to include some with open, shallow flowers for easier access to nectar and pollen and/or extrafloral nectaries. Extrafloral nectaries are glands found on leaf surfaces and margins, petioles, leaf and flower bracts, and sepals that provide natural enemies with sugar, water, and amino acids.

Insectary habitats also can provide areas for natural enemies to seek refuge from the sun, hide from predators, and find cover in which to spend the winter. By offering plants that vary in height and growth form, and by allowing the dead vegetation to remain throughout winter, you increase your chances of attracting biological control insects to your garden.

ANNUAL PLANTS

So, what plants create the most effective insectary? Many studies have focused on the value of annual plants for conservation biological control because they are often prolific pollen and nectar producers, generally have low seed or plant cost, and can be incorporated directly

△ There are many ways to incorporate insectary plants into your home landscape. Containers are a great way to get creative with small, mobile insectaries using annual plants. Add container insectaries to patio gardens or even place them directly adjacent to or in vegetable gardens. Move containers throughout the season to the locations in your landscape where they are needed most.

△ Looking to create a larger insectary? Buffer strips of annual or perennial plants can be established in or adjacent to vegetable gardens or home orchards. Rows of annuals also work great intermixed with vegetables. If you are interested in creating an insectary with perennial plants, using native species is a great option. These plants can be added throughout existing perennial beds, between rows of tree fruit, or placed together in a garden bed to create a unique and highly attractive native plant insectary for bees and natural enemies.

within fruit and vegetable crop beds each spring. Several annuals have been identified as highly attractive to beneficial arthropods, including borage (*Borago officinalis*), buckwheat (*Fagopyrum esculentum*), coriander or cilantro (*Coriandrum sativum*), dill (*Anethum graveolens*), fava bean (*Vicia faba*), phacelia (*Phacelia tanacetifolia*), and sweet alyssum (*Lobularia maritima*).

Annual Attractors: The Sweet Seven

Many annual plants are prolific nectar and pollen producers, including those shown here, which have a proven track record for attracting natural enemies. When designing an insectary, these plants can be an important component, but it's not ideal to rely solely on one or two species of insectary plant. Like us, natural enemies vary in their preferences because different plants may support different prey or provide nectar and pollen at different times of year. So, selecting a diversity of species is the best strategy to sustain an abundant and diverse community of predators and parasitoids.

Dill
(*Anethum graveolens*)

Borage
(*Borago officinalis*)

Coriander or Cilantro
(*Coriandrum sativum*)

Buckwheat
(*Fagopyrum esculentum*)

Sweet Alyssum
(*Lobularia maritima*)

Phacelia
(*Phacelia tanacetifolia*)

Fava Bean
(*Vicia faba*)

Annual plants work well in containers, as mobile insectaries that can be moved throughout a home landscape. Annuals also can be planted in rows between vegetable and fruit crops or as borders surrounding food gardens. Sweet alyssum is a great choice to add nectar and pollen resources directly within vegetable gardens because it is relatively low growing and blooms throughout the growing season.

Beyond the "sweet seven," many other annual plants can provide abundant resources for natural enemies. The best way to optimize an insectary planting is to experiment with a diversity of flowering species in containers and beds. Observation will illustrate which plants are most favored by bees and natural enemies. Do a little research, however. Some hybrids produce little-to-no nectar and pollen, so these plants will not provide much support to natural enemies.

EXTRAFLORAL NECTARIES

The presence of extrafloral nectaries – glands found on leaf surfaces and margins, petioles, leaf and flower bracts, and sepals that provide nourishment to natural enemies – is another plant feature to consider when designing your insectary. Plants with these nectaries produce nectar not only in flowers, but also on other plant parts. Many species including ants (Formicidae), lady beetles (Coccinellidae), predatory and parasitoid wasps (Hymenoptera), and spiders (Araneae) are known to feed from extrafloral nectaries. More than 100 plant families contain species that have extrafloral nectaries, including many found in home landscapes. You may already be growing some of these crop and ornamental

Photo: MaLisa Spring

△ *Nectar isn't only found in flowers. Many plants have extrafloral nectaries that provide natural enemies with an accessible source of sugar, amino acids, and water. These glands can be found on many different plant parts. Northern caltalpa (Catalpa speciosa), for example, has extrafloral nectary glands along its leaf midrib.*

Photo: Thomas Wilson

△ *Here a predatory wasp was photographed consuming extrafloral nectar from a common vetch plant (Vicia sativa) at Herring Run Park in Baltimore City, Maryland. Attracting predators and parasitoids with this sugary-rich food can help to protect these plants from damaging herbivores.*

plants now. The addition of others can help to improve the abundance and diversity of natural enemies in your garden; be sure to select species that are suitable for your location and are not invasive.

Extrafloral Nectaries

Extrafloral nectar is thought to be produced as a means of plant defense because these nectar sources specifically attract natural enemies that consume plant-feeding insects. This table illustrates the diversity of trees, shrubs, garden plants, crops, and weeds found in home landscapes that produce extrafloral nectar.

Abutilon	(Indian mallow and velvet leaf)
Asclepias	(milkweed)
Catalpa	(catalpa)
Cereus	(night-blooming cacti)
Chamaecrista fasciculata	(partridge pea)
Circaea	(enchanter's nightshade)
Crataegus	(hawthorn)
Cucurbita	(pumpkin and squash)
Euphorbia pulcherrima	(poinsettia)
Ferocactus	(barrel cacti)
Fritillaria	(fritillaries and mission bells)
Gossypium	(cotton)
Helianthus	(sunflower)
Hibiscus	(hibiscus)
Impatiens capensis	(jewelweed)
Ipomoea	(morning glory and sweet potato)
Malus	(apple)
Opuntia	(prickly pear)
Pachycereus	(organ pipe and totem cacti)
Paeonia	(peony)
Passiflora	(passionflower)
Phaseolus	(fava bean and lima bean)
Populus	(poplar)
Prunus	(plum, cherry, peach, nectarine, apricot, and almond)
Pteridium aquilinum	(bracken fern)
Pyrus	(pear)
Ricinus communis	(castor bean)
Robinia	(locust)
Salix	(willow)
Sambucus	(elderberry)
Sansevieria	(snake plant)
Silene	(champion and catchfly)
Smilax	(greenbrier)
Sporobolus heterolepis	(prairie dropseed)
Syringa	(lilac)
Viburnum	(viburnum)
Vicia sativa	(common vetch)
Vigna unguiculata	(cowpea)

Native Plants for Natural Enemies

Gardening with native plants is a great way to enhance biodiversity in your home landscape. Visit a reputable garden center or native plant producer to design a native plant insectary specific for the soil, sun, and moisture conditions present in your home landscape. Many native plants have also been shown to be highly attractive to natural enemies, including those shown here.

Giant Hyssop
(Agastache)

Photo: Shutterstock.com

Lobelia
(Lobelia)

Photo: Shutterstock.com

Milkweed
(Asclepias)

Photo: Shutterstock.com

Asters
(Aster)

Photo: IS Image

Rabbitbrush
(Chrysothamnus)

Photo: Shutterstock.com

Prairie Clover
(Dalea)

Photo: Shutterstock.com

Coneflower
(Echinacea)

Photo: Shutterstock.com

Joe Pye Weed
(Eupatorium)

Photo: Shutterstock.com

Sunflower
(Helianthus)

Photo: Shutterstock.com

Blazing Star
(Liatris)

Photo: Shutterstock.com

Lupine
(Lupinus)

Photo: Shutterstock.com

Bee Balm
(Monarda)

Photo: Shutterstock.com

Beardtongue
(Penstemon)

Mountain Mint
(Pycnanthemum)

Rhododendron
(Rhododendron)

Sage
(Salvia)

Cup Plant
(Silphium)

Goldenrod
(Solidago)

Meadowsweet
(Spiraea)

Snowberry
(Symphoricarpos)

Aster
(Symphyotrichum)

Spiderwort
(Tradescantia)

Ironweed
(Vernonia)

Culver's Root
(Veronicastrum)

NATIVE PERENNIALS

Many gardeners are interested in incorporating native flora into their home landscape, and these perennials are excellent additions to an insectary planting. Perennials offer pollen and nectar resources, alternative prey, and cover for natural enemies to overwinter if they are not cut back in the fall. Native plants are also adapted to the conditions present in your region and often require less watering and fertilizer than non-natives. As an added side benefit, many of these plants have become rare or even endangered, so incorporating them into your home landscape improves native biodiversity in your area. When selecting your plant mix, it is a good idea to talk with a local native plant producer to learn more about the native flora for your region and to determine the species that will do best. These plantings can include native trees, shrubs, and herbaceous plants.

There are excellent resources online that offer lists of plants known to be attractive to natural enemies and pollinators. Michigan State University researchers have developed a listing of recommended native plants to attract both bees and natural enemies that is suitable for use in the Great Lakes region (nativeplants.msu.edu), and The Xerces Society for Invertebrate Conservation has developed regional plant lists covering the United States to support pollinator populations (xerces.org). These plants were selected with bees in mind, but many will be highly attractive to natural enemies as well.

△ *Get creative with your lawn! Companies such as American Meadows offer alternative lawn seed mixes that include fescue grasses and low-growing, flowering plants.*

Photo: American Meadows

Lawns as a Habitat

Looking for other ways to create beneficial insect habitat? Lawns take up the greatest area of most residential landscapes, but typically the goal is to keep these weed free. Making peace with weeds saves on herbicide use and provides additional resources for beneficial insects. But if you are not ready to surrender in the battle against dandelions, there are also "no mow" grass seed mixes available that contain annual and perennial flowering plants such as clover, yarrow, sweet alyssum, thyme, poppies, and daisies.

Shelter from the Elements

Another way to attract natural enemies to your yard is to incorporate garden features with protected cracks and crevices in which beneficial insects can take shelter. For example, natural enemies will nestle under rocks and in woodpiles. Adding or maintaining these elements provides a habitat for predators and parasitoids during the growing season and a protected space to spend the winter.

You can also add nesting boxes and retreats to a garden to provide protected spaces for natural enemies to nest and/or overwinter. Many are actually designed for bees, but they can also be colonized by solitary wasps. The bee nesting boxes available online or from garden centers are often designed to attract mason bees (Megachilidae: *Osmia*) and have 16 mm holes. To attract a greater diversity of bees and wasps, you can embark on a relatively easy DIY project. Take a 4 x 4 inch (10 x 10 cm) or 4 x 6 inch (10 x 15 cm) block of wood and drill holes of varying diameter into it (2.5 to 10 mm), with a depth of 3 to 5 inches (8 to 13 cm) for holes smaller than 6 mm, and 4¾ to 5½ inches (12 to 14 cm) for larger holes (recommended by the Xerces Society for Invertebrate Conservation). You can paint the outside of the house, add a roof, and mount the block to a post, fence, or shed where it will be secure and ideally catch the morning sun. Another option is to collect hollow stems from plants in your home landscape and place them together inside a structure. Paper straws are commercially available that will be readily colonized as well. You can leave these stems or straws outside in their structure during the winter, but you are likely to increase survivorship if you place them in a plastic container with a mesh vent and store them in a

△ *Many garden centers and online businesses sell nesting blocks designed for mason bees (Osmia). This one has 16 mm holes, which can be lined with papers straws. In the fall, remove the straws with developing bees and wasps inside and store them in a cool place for the winter. In the spring, allow the pollinators and natural enemies to emerge from a vented container out of direct sunlight and protected from rain.*

△ *Nesting sites for bees and wasps can range from simple to complex. This "Bee Hotel" at Michigan State University incorporates nesting blocks with holes of varying sizes, paper straws, and hollow stems to encourage nesting by a wide diversity of bees and wasps.*

refrigerator or unheated building. In the early spring, place the plastic container in the garden and remove the mesh to allow emerging bees and wasps to leave.

IMPORTANT NOTE: *The Xerces Society recommends disinfecting your nesting blocks to prevent buildup of pathogens and parasites. You can line nesting blocks designed for mason bees (16 mm holes) with commercially available paper straws to extend the life of the box. Remove the straws in the fall to overwinter and disinfect the block by soaking it in a 30 percent bleach solution. Replace unlined blocks after two seasons; you can allow emergence from the old block by placing it in an emergence container.*

Other types of overwintering retreats are also commercially available, including some that are marketed specifically for natural enemies, such as lady beetle (Coccinellidae) and lacewing (Chrysopidae) shelters. They are typically open boxes with small slits for the natural enemies to enter. The idea is that these shelters provide protection from the elements and potential predators of the natural enemies during the growing season as well as a protected place to overwinter. I personally have not tried any of these to see if arthropods will colonize them, but DIY instructions are readily available online, so it might be a fun rainy day project to make one and give it a try.

Mulching

Adding mulch to the garden is a great conservation biological control strategy to enhance ground-dwelling predators such as wolf spiders (Lycosidae) and ground beetles (Carabidae). Spreading mulches, such as straw, bark, or woodchips, between plants adds moisture and

△ *Natural enemy shelters such as this one offer a protected space to escape predators and unfavorable temperatures. Anthony Anfuso designed and built this lady beetle house. He discovered, however, that building a lady beetle (Coccinellidae) house does not guarantee it will be used by lady beetles; this house ended up being colonized by wasps. So you may not attract your target, but there is a good chance that natural enemies will investigate this garden addition!*

Photo: Anthony Anfuso

offers natural enemies sun protection, enabling them to effectively move throughout the garden without drying out. Mulches may also support alternative prey for predators. For example, soil-dwelling mites can be enhanced with the addition of mulch. These mites are not garden pests and can serve as alternative prey for predatory mites, which can build up in mulched gardens and suppress spider mite populations. Keep in mind, however, that mulches are not uniformly beneficial for arthropods. Some soil-nesting bees prefer bare soil, so leaving some spaces without mulch might encourage local nesting by these species within your garden.

In addition, if you are looking for a break from weeding, the presence of weeds can provide natural enemies with shelter, alternative prey, and in some cases, pollen and nectar resources. Other plants can also be established within gardens as living mulch. Plants such as hairy vetch (*Vicia villosa*), rye grass (*Lolium*), or alfalfa (*Medicago sativa*) can be seeded into a garden space in the fall. In the spring, strips can be tilled into this mulch, and crops may be established into the strips. The strips of mulch remaining between garden rows provide shelter and alternative prey for natural enemies.

Natural Enemy Food Sprays

Commercially available food sprays can be applied directly to plant foliage to attract and retain natural enemies. These products, which contain sugar and sometimes have added protein resources, can be an easy and effective way to provide food for natural enemies when prey populations are low. You can also make your own sugar solution by heating one pound (450 g) of sugar mixed with one gallon (3.8 L)of water and stirring until all the sugar is dissolved. The solution can then be cooled and applied with a hand-held pump sprayer.

3

Monitoring and Collecting Good Garden Bugs

With all of the interesting natural enemies foraging within your home garden, it's helpful to have a few simple tools on hand that allow you to take a closer look. Carefully inspecting plants is the easiest way to scan your home landscape for natural enemies at work. If you want to find the greatest diversity of species, inspect your plants at various times throughout the day and evening, keeping in mind that you are likely to find less activity during the hottest hours. In addition to looking at open leaves and stems, look for possible hiding spots natural enemies might use, such as the shelter of a leaf that has not yet unfurled or underneath a developing squash fruit, nestled in straw mulch. It is also always a good idea to have a camera along so you can quickly snap photos for identification purposes. Entomologists are usually able to give you some information about a mystery insect — even from a blurry cell phone photo!

Tools of the Trade

Entomologists use many different tools for insect collecting. These tools can be made from everyday household items or purchased relatively inexpensively. Using the methods described here, you can observe your backyard insects while they are still living before releasing them back into your garden space.

Netting Above-Ground Natural Enemies

Insect nets can be a great way to find out just how many insects are in a section of tall grass against your fence or to gain an up-close view of a fast-moving wasp or fly. There are two main types of insect nets: sweep nets and aerial nets.

A sweep net, which is designed for sweeping through tall vegetation, has a canvas bag attached to the end of it for collecting insects from lawns and meadows. You can also hold a sweep net under the branches of trees or bushes and use a stick to beat the foliage, dislodging the insects into your net. To inspect your sweep net catch, shake the net vigorously to knock the contents into the bottom; then invert it and dump the catch onto a white plastic tray or larger light-colored sheet of poster board or cardboard. The insects will begin to fly and crawl away quickly, so if you want more time to inspect your catch without killing them, dump the contents of the net into a plastic storage bag, zip it closed, and place it in the refrigerator for fifteen minutes to slow down the bugs first.

Photo: Mary M. Gardiner

△ Canvas sweep nets are the perfect tool to help learn more about the diversity of natural enemies foraging within meadows and lawns. Research scientist Chelsea Smith is looking for native lady beetles in a planted prairie.

Aerial nets consist of a mesh bag and are designed for collecting individual insects off of flowers or in midair. These nets will allow you to see the captured insects through the mesh, making it easier to place it into a container for observation. Consider buying a net through a company that sells the handle and net bags separately, allowing you to use both a sweep and aerial bag with the same handle.

△ *A mesh aerial net is best to catch flying arthropods such as butterflies and wasps. Graduate student Scott Prajzner is using an aerial net to catch and identify pollinators in an urban garden. Once an insect is trapped within an aerial net, the mesh fabric makes it much easier to transfer it to a container for closer inspection.*

△ *After sweeping the vegetation with a canvas net, empty the contents of the net onto a white tray or piece of light colored poster board or cardboard. Research scientist Nicole Hoekstra and graduate student MaLisa Spring are inspecting a sweep sample collected from a Cleveland vacant lot habitat.*

△ *On a trip to Texas, graduate students Andrea Kautz and Erin O'Brien carefully transfer a wasp from their net to a collection container to get a closer look.*

Trapping Ground-Dwelling Natural Enemies

There are so many interesting natural enemies that forage on the soil surface, but many may be difficult to catch because they are fast moving, a little creepy, and tend to be active at night. Setting one or more pitfall traps in your garden is a great way to get a closer look at these elusive creatures. A pitfall trap can be made simply by digging a hole to fit a plastic cup. Unsuspecting natural enemies will stumble into it and become trapped. Often, pitfall traps are filled with water or other liquids to kill and preserve the specimens, but a dry pitfall is a great way to monitor activity without harming your natural enemies. When digging a hole for your trap, be sure to make it deep enough so that the top of the cup is even with or slightly below the soil surface. Check the weather before setting the trap to confirm that there is no rain in the forecast for at least twenty-four hours. Then set the trap in the morning and check it throughout the day so you're able to monitor a variety of insects active at different times of day and to ensure the first catch doesn't eat any of the insects caught later. If left overnight, you are likely to find a large catch of spiders (Araneae) and ground beetles (Carabidae) among other nocturnally-active arthropods. If you want to compare how different management practices, such as putting down different types of mulch, may affect the abundance and diversity of these predators in your backyard, sink pitfalls in different garden spaces and compare your results.

Photo: MaLisa Spring

△ *Use this trick when setting pitfall traps: Place two cups in the hole. When you remove the top cup to inspect your catch, the cup can be easily put back into place. This trap is monitoring ground-dwelling arthropods within a rain garden. This cup has soapy water to collect spiders for a research study. To catch and release live, the same method can be used without the soap and water.*

4

Mantids (Mantodea)

Considered one of the most charismatic garden predators, mantids have inspired everything from poems and paintings to the fighting strategies used by martial artists. They are often called praying mantids because most species are sit-and-wait predators that hold their front legs together as if in prayer when waiting for prey to come into reach. If a potential meal does come near, the mantid will grasp it with its front legs before killing and consuming it with its strong mandibles.

Mantids are in the order Mantodea, and all are in the arthropod family Mantidae. They are large, measuring from 2 to 4½ inches (5 to 11.5 cm) in length, with slender legs and long, thin bodies. Their front legs are modified for grasping prey, and spines line their femur and tibia. Mantids have a triangular-shaped head with two large compound eyes. They have amazing articulation of their head, which aids in their ability to hunt and allows them to quickly spot potential predators, such as birds. Mantids eat a wide variety of prey—both pests and other beneficial species—including most insects and spiders. Larger mantid species are even known to eat small reptiles and amphibians on occasion!

Although they are large insects, mantids blend in well with their surroundings, often resembling leaves or sticks. They also have a characteristic rocking or swaying behavior, which may be another adaptation for camouflage, to appear as if they are foliage moving in the wind, but it is also thought to help the mantid better separate potential prey from a background of vegetation. Many mantid species can fly, but typically, it is the males that do most of the flying. Females often do not fly or fly infrequently. They attract males for mating by releasing pheromones—very specific volatile odors that help males to locate mates of their species.

Mantids express interesting courtship behavior, with males of some species performing an elaborate dance prior to mating. This is believed to be a way for the male to overcome the female's tendency to grab and eat him. The frequency of females eating their mates may be exaggerated because it is based primarily on observations made in laboratories, where the insects may not act like they do under natural conditions. Females may, however, kill the male after mating or behead him during mating. Consumption of the male can be advantageous because the female will produce more eggs with this large meal, and the male will die soon after mating anyway. Females deposit their eggs in a mass of foam that hardens into a large egg case called an ootheca, which can contain up to 400 eggs. In some mantid species, the female will guard her eggs until they hatch. Mantids overwinter in the egg stage and develop by gradual metamorphosis. This means that in the spring, wingless nymphs will hatch from the egg case and develop through several molts before reaching the adult stage.

There are several species of exotic and native mantids in the United States. Native mantids are found primarily in the southern and western United States.

Nonnative Mantids

In the northern United States, two introduced, exotic species, the European mantid (*Mantis religiosa*) and the Chinese mantid (*Tenodera sinensis*), are the only species commonly found. Both of these species were introduced deliberately to provide pest control in agricultural crops. Although they are voracious predators of other arthropods, mantids pose no danger to humans because they do not bite or sting.

European Mantid (*Mantis religiosa*)

The European mantid is a nonnative species that was introduced into the United States in the late 1800s and is found throughout the country but is least common in very hot and humid or very dry regions. Measuring approximately 2¾ inches (7 cm) when fully grown, the European mantid exists in both green and brown color forms. This species can be identified by a bull's-eye spot present on the inside of the foreleg. They have become so common that, despite being a nonnative species, the European mantid is Connecticut's state insect.

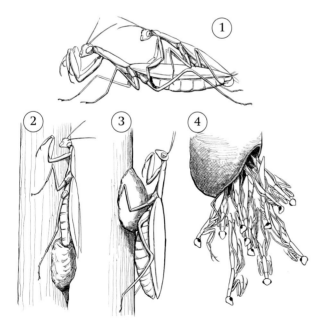

1. *Male mantids often perform a courtship dance to engage their prospective mate. Mating may end with the female eating the male, but the frequency of this may be exaggerated.*

2. *Females deposit their eggs in a mass of foam that hardens. This egg mass, which looks a bit like Styrofoam, is called an ootheca.*

3. *In some species, the female will guard the ootheca until the nymphs hatch.*

4. *An ootheca can contain hundreds of eggs, but not all of the nymphs that emerge will survive, due in part to significant sibling cannibalism.*

◁ *The European mantid (Mantis religiosa) is a very common nonnative species. To identify this species, look for a bull's-eye spot on the inside of each raptorial (grasping) front leg.*

Photo: Joyce Gross

Chinese Mantid (*Tenodera sinensis*)

The Chinese mantid, which measures 4½ inches (11.5 cm) long when fully grown, is the largest species found in the United States. This species was introduced in the late 1800s and is now very commonly found in the northern United States. They are pale green or tan with a green lateral strip along the edge of their front wings. Egg cases of the Chinese mantid are available commercially for release in home gardens. Mantids are extreme generalist predators, feeding on larger pests such as grasshoppers but also on other beneficial insects. Releasing mantids is unlikely to lead to suppression of most common garden pests because they generally will not take much interest in small prey such as aphids (Aphididae) or spider mites (Tetranychidae). However, it is fascinating to watch them forage, and they are a great way to interest children in entomology!

△ *The Chinese mantid (Tenodera sinensis) is a nonnative species in the United States. They have a green lateral stripe along the edge of each front wing.*

Photo: D. Shetlar

Brunner's Mantid (*Brunneria borealis*)

The Brunner's mantid is native to the southeastern United States. Also called the Brunner's stick mantid, this species is named for its very slender body, which is 2½ to 3½ inches (6.5 to 9 cm) in length when fully grown. They have greatly reduced wings and are flightless. Interestingly, this mantid reproduces through parthenogenesis, or asexual reproduction, where embryos grow and develop without fertilization. Given their reproduction, there are only female Brunner's mantids, no males. The females produce a very distinctive ootheca that has a hornlike projection on one end. Like all mantids, they can feed on a large variety of insects, but the Brunner's mantid prefers those in the insect order Orthoptera, such as grasshoppers (Acrididae) and crickets (Gryllidae).

Grizzled Mantid (*Gonatista grisea*)

The Grizzled mantid is native to the southeastern United States and measures 1¼ to 1½ inches (3 to 4 cm) in length. It has a mottled appearance, which provides excellent camouflage against tree bark and lichens. The Grizzled mantid has a broad and flattened body. Females have short wings that do not cover the abdomen, but they do in males.

Photo: Ken Carman, head naturalist Roxbury Park, South Carolina.

◁ *The Brunner's mantid (Brunneria borealis) has a very slender body. Only females exist of this flightless species because they reproduce asexually, without mating.*

△ *The grizzled mantid (Gonatista grisea) has a mottled brown and green appearance, which provides excellent camouflage because they hunt on tree bark for prey.*

Ground Mantids (*Litaneutria minor* and *Litaneutria obscura*)

Ground mantids are unique in that instead of adopting the typical sit-and-wait predatory strategy of most mantids, these active hunters stalk their prey on the ground. There are two species, *Litaneutria minor* and *Litaneutria obscura*, both of which are native to the western United States. *Litaneutria minor* is commonly called the agile ground mantid because they can be found running swiftly along the ground in search of prey. They are found in the desert southwest, eastern California, Oregon, and Washington and are ¾ to 1¼ inches (2 to 3 cm) in length. They are also found in southwestern Canada and are the

◁ *Unlike other mantids, ground mantids (Litaneutria minor) run after insect prey; they are not sit-and-wait hunters.*

◁ *The little Yucatan mantid (Mantoida maya) is dark in color with large eyes and a short, square-shaped pronotum.*

only native Canadian mantid. *Litaneutria obscura* is commonly called the obscure ground mantid and is found in Arizona, California, Texas, and New Mexico. It also measures ¾ to 1¼ inches (2 to 3 cm) long. Both species are cryptically colored, with a brown and gray mottled appearance. Both females and males have very short wings as adults and do not fly.

Little Yucatan Mantid (*Mantoida maya*)

This species is native to Florida and is also found in Mexico. The little Yucatan mantid is small for a mantid, measuring ½ to ¾ of an inch (1.5 to 2 cm) in length. The nymphs are ant or wasp mimics. They have a red head and thorax and a black abdomen. Adults of this species look the least like a typical praying mantid. They are dark in color, with reddish-brown bodies and wings. They have a short, square-shaped pronotum instead of the long, slender one typical of other mantids.

Unicorn Mantids (*Pseudovates arizonae* and *Phyllovates chlorophaea*)

The unicorn mantid takes its name from the projection

△ *Both the Texas and Arizona unicorn mantid have two projections on their head that form what looks like a single horn, leading to their common name.*

on its head that resembles a unicorn horn. A closer look reveals that it is actually two adjacent horns. The Arizona unicorn mantid (*Pseudovates arizonae*) measures 2½ to 2¾ inches (6.5 to 7 cm) and is found in south-central and southeastern Arizona. Its body is dark brown with light brown stripes, and it has light green leaflike forewings. The legs are light with dark brown banding. The Texas unicorn mantid *(Phyllovates chlorophaea)* is 2 to 2¾ inches (5 to 7 cm) and found in the southern part of the state. The adults have brown bodies with light banding on their legs and bright green leaflike wings.

California Mantid (*Stagmomantis californica*)

The California mantid is native to the western United States and is commonly found in arid desert regions within this range. They are 2 to 2½ inches (5 to 6.5 cm) in length and may be green, yellow, or brown. Females have short wings that do not cover their abdomen, and males have longer wings that do, but both sexes are able to fly. The forewings are mottled or suffused with dark brown or black. They have dark purple hind wings. Males have dark bands along the top of their abdomen. They commonly hunt in trees and shrubs.

△ *The Carolina mantid* (Stagmomantis carolina) *can adjust its color to match its environment. In females, the wings extend three-quarters the length of the abdomen and have a black patch.*

Carolina Mantid (*Stagmomantis carolina*)

The Carolina mantid is native to the United States, and it measures 2 to 2½ inches (5 to 6.5 cm) in length as an adult. It is found from New York south to Florida and west to Utah, Arizona, and Texas. The Carolina mantid is gray, brown, or green and can adjust to blend in with its surroundings by molting to change color. In females, the wings only extend three-quarters of the length of the abdomen and have a black patch. The abdomen of the female is strongly widened in the middle. The Carolina mantid is the state insect of South Carolina.

5

True Bug Predators (Hemiptera)

Do you enjoy a morning smoothie? How about eating *all* your meals through a straw? This is the reality for insects in the order Hemiptera, which consume their food through a strawlike mouthpart called a stylet or beak. Predatory true bugs use this mouthpart to pierce their insect victims, releasing saliva that contains enzymes to partially digest their meal before the predator consumes it. Because of this, some true bug predators can inflict a painful bite.

In addition to their strawlike mouthpart, predatory Hemiptera can be identified by the presence of a thickened forewing that is leathery at its base where the wing attaches to the thorax and membranous at its tip. Their hind wings are completely membranous. This was the inspiration for the scientific name Hemiptera, meaning "half-wing."

Insects in the order Hemiptera develop by gradual metamorphosis, meaning that females lay eggs that hatch into wingless nymphs and develop through several molts before reaching the adult stage. As both nymphs and adults, predatory true bugs feed on a wide range of garden pests. In addition to predators, this large and diverse order includes many insects that feed on plant sap. Some of these, such as aphids (Aphididae), stink bugs (Pentatomidae), and scales (Coccoidea), are significant

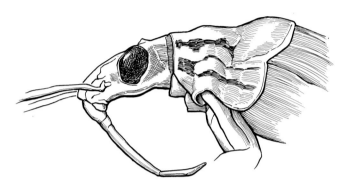

△ *Predatory true bugs consume their prey though a strawlike mouthpart called a stylet or beak. They release saliva containing digestive enzymes into their prey that allow them to consume a liquid diet.*

home garden pests. This chapter identifies some of the common families of beneficial predatory Hemiptera you may encounter in your home garden and backyard water features.

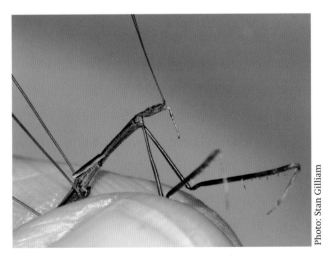

Photo: Stan Gilliam

△ *It's clear this thread-legged bug is none-too-pleased about being held. The captive insect is stretching its head and directing its piercing beak toward that thumb! Take care if you handle a true bug predator because some can inflict a painful bite.*

Flower Bugs or Minute Pirate Bugs (Anthocoridae)

The family Anthocoridae is commonly referred to as flower bugs or minute pirate bugs. As you may have guessed from the name, these insects are tiny and will not be easy to spot in the garden, but they are a common and important group of predators that can be found throughout the United States. Adult minute pirate bugs measure 3 to 6 mm long and are oval to triangular in shape. They typically have a dark head and thorax and forewings with light and dark patches. Nymphs are teardrop shaped and bright yellow to pink. The nymphs are very fast moving. Two of the most common genera of Anthocoridae found in the garden are *Orius* and *Anthocoris*.

Female minute pirate bugs lay tiny eggs embedded within plant tissues. Nymphs will hatch and begin foraging for food. They undergo several molts before reaching the adult stage. The adults are active within the garden for about one month, and females can lay more than 100 eggs over their lifetime.

Both nymphs and adults feed on a diversity of garden pests, including aphids (Aphididae), scales (Coccoidea), spider mites (Tetranychidae), thrips (Thripidae), small caterpillars (Lepidoptera), and insect eggs. These predators also feed on pollen and nectar from flowering plants, providing these resources can increase their reproductive potential and longevity. They are able to bite humans, but their bite is only mildly irritating.

Photo: Scott Justis

△ *An adult minute pirate bug in the genus* Orius. *Note the light and dark patches on the forewings.*

Photo: Charley Eiseman

▷ *A minute pirate bug nymph* (Anthocoridae). *Watch for a fast-moving insect hunting aphids and other soft-bodied plant pests. They are also very commonly found in flowers, feeding on pollen and nectar.*

Big-Eyed Bugs (Geocoridae)

Photo: Stephen Cresswell

△ Note the very prominent eyes of Geocoris punctipes.

Adult big-eyed bugs are small predators (3 to 5 mm in length) that are brown or black with very prominent eyes. Nymphs look similar to adults but without fully formed wings. Females deposit eggs either on plants or in soil, and nymphs undergo several molts before reaching the adult stage. Big-eyed bugs overwinter as adults, and several generations are possible within a single growing season. They feed on a diversity of prey, including insect eggs, caterpillars (Lepidoptera), leafhoppers (Cicadellidae), spider mites (Tetranychidae), thrips (Thripidae), whiteflies (Aleyrodidae), and aphids (Aphididae). The most common big-eyed bug predator found in gardens is *Geocoris punctipes*, which can be found throughout the United States.

Damsel Bugs (Nabidae)

All species in the family Nabidae are predatory. The majority are a dull tan with slightly enlarged raptorial front legs and a curved beak. Adult damsel bugs are approximately 6 to 10 mm in length. Adults and nymphs look similar, but nymphs will have wing buds instead of fully formed wings. Wings will develop as they molt and become functional during the final molt into the adult stage.

Damsel bugs complete one to five generations per year depending on location, and most species overwinter as adults in leaf litter. Female damsel bugs begin laying eggs in the early spring shortly after they emerge from overwintering. They insert the eggs, which are elongated and flattened, into plant tissue. Eggs hatch into nymphs that feed on arthropod prey and complete several molts to reach the adult stage.

Nabidae use their enlarged front legs to capture and hold prey. They are important garden predators that feed on insect eggs, aphids (Aphididae), leafhoppers (Cicadellidae), small sawfly larvae (Hymenoptera, Symphyta), spider mites (Tetranychidae), and small caterpillars such as corn earworm (Noctuidae: *Helicoverpa zea*), European corn borer (Crambidae: *Ostrinia nubilalis*), imported cabbageworm (Pieridae: *Pieris rapae*), and armyworms (Noctuidae).

Photo: Stephen Cresswell

△ The black damsel bug Nabis subcoleoptratus *is wingless as an adult. Look for the light strip around its abdomen to aid in identification.*

This family includes common garden predators in the genus *Nabis* including *Nabis americoferus* (broad distribution in North America), *Nabis roseipennis* (found in the north central United States) and *Nabis alternatus* (common in the western United States). The Nabidae family also includes the black damsel bug *Nabis subcoleoptratus*, which is wingless as an adult, black with light yellowish legs, and has a light stripe around the exterior of its abdomen.

△ *Damsel bugs in the genus* Nabis *are common garden predators. Most are tan with enlarged front legs for grasping prey. Nymphs look very similar to adults but without fully formed wings.*

Predatory Stink Bugs (Pentatomidae)

The stink bug family Pentatomidae includes many plant-feeding pests but also a few key predatory species. These predators can be identified by their five-sided shield-shaped body and large triangular plate called a scutellum, which is part of their thorax. Predatory stink bug nymphs are more rounded than shield-shaped and are often brightly colored or patterned. Predatory stink bugs overwinter as adults or late instar nymphs, and in the spring, females begin laying clusters of barrel-shaped eggs on plant leaves and stems. Nymphs undergo several molts to reach the adult stage.

Spined Soldier Bug (*Podisus maculiventris*)

The spined soldier bug is a common predatory stink bug found in gardens throughout the United States. These predators are about ½ an inch (1 cm) in length and brown, with "pointy shoulders" or points on either edge of their

thorax. Spined soldier bugs also have a black spot on the membranous tip of each front wing. They feed on a diversity of garden pests, including caterpillars (Lepidoptera) and beetle larvae (Coleoptera).

△ *The spined soldier bug* (Podisus maculiventris) *has "pointy shoulders" and a black spot on the tips of its wings.*

△ *Two-spotted stink bugs (Perillus bioculatus) are black with light outlining on the thorax and scutellum.*

Photo: Thomas Bentley (thomasbentley.com)

Photo: Gayle and Jeanell Strickland

△ *Seems like a nicer common name could be found for the giant strong-nosed stink bug (Alcaeorrhynchus grandis) because it is actually an attractive species with red, orange, and yellow coloration mixed with black mottling. It also can be identified by the presence of two points on either side of its pronotum.*

Photo: Christy Beal

△ *The Florida predatory stink bug (Euthyrhynchus floridanus) has a deep blue body with three orange markings on its scutellum and additional coloration on the edges of its abdomen.*

Two-Spotted Stink Bug (*Perillus bioculatus*)

The two-spotted stink bug is 8 mm in length and black with yellow, orange, or red outlining the abdomen and thorax. Commonly known to feed on Colorado potato beetle eggs and larvae (Chrysomelidae: *Leptinotarsa decemlineata*), this species is found throughout the United States.

Florida Predatory Stink Bug (*Euthyrhynchus floridanus*)

The Florida predatory stink bug is found in the southeastern United States. This species ranges in length from ½ to ¾ of an inch (1 to 2 cm), and is dark blue to black, with three light-colored spots outlining the scutellum. The Florida predatory stink bug feeds on caterpillars (Lepidoptera), Colorado potato beetles (Chrysomedlidae: *Leptinotarsa decenilineata*), other beetle larvae (Coleoptera), leafhoppers (Cicadellidae), and other pest stink bugs.

Giant Strong-Nosed Stink Bug (*Alcaeorrhynchus grandis*)

As the name suggests, the giant strong-nosed stink bug is a large predator, measuring ¾ of an inch (2 cm) in length. This stink bug is mottled light and dark brown and has two points on either edge of its pronotum — the top surface of the insect's prothorax. They feed predominately on caterpillars (Lepidoptera) and are found in the southern United States.

Anchor Stink Bug (*Stiretrus anchorago*)

The anchor stink bug is 7 to 9 mm in length with a long, rounded scutellum. This species can vary from all black to black with yellow to red markings. The anchor stink bug is found in the eastern and southern United States and feeds predominately on caterpillars (Lepidoptera) and beetle larvae (Coleoptera).

Photo: Brandon Woo

△ *A key feature of the anchor stink bug* (Stiretrus anchorago) *is its long and rounded scutellum. The coloring on adults is variable but in general they have dark coloring running down the center of their back with lighter margins containing dark spots.*

Thread-Legged Bugs, Ambush Bugs, and Assassin Bugs (Reduviidae)

Reduviidae is one of the largest families of Hemiptera, and all species within this group are predatory or blood-feeding. They are sit-and-wait predators that grasp prey with raptorial front legs and subdue it by using their curved beak to inject a paralytic fluid. Reduviidae also use their curved beak to ingest bodily fluids from prey. These predators usually overwinter as adults, but some species do overwinter as eggs or nymphs. They may have one or two generations per year. Females lay eggs in clusters, and after the offspring hatch, they undergo four to seven molts before reaching adulthood, depending on the species. There are many species of Reduviidae that can be found in home gardens, including the thread-legged bugs, ambush bugs, and assassin bugs.

Thread-Legged Bugs

As their name suggests, thread-legged bugs have very slender bodies and long, thin legs. Adults are commonly 3 to 10 mm, but may be up to 1½ inches (4 cm) in length. The nymphs of thread-legged bugs look similar to adults, but without fully formed wings. These bugs walk on the hind two pairs of legs and use the front raptorial (grasping) pair to capture prey. It can be easy to confuse thread-legged bugs with walking sticks, which feed on plants and have chewing mouthparts. If you come across what appears to be a thread-legged bug, look closely for their piercing beak. Many genera of thread-legged predators are found in the United States, including *Barce* (½ to ¾ of an inch [1.5 to 2 cm]), *Pseudometapterus* (4 to 20 mm), *Empicoris* (4 to 8 mm), and *Emesaya* (1¼ to 1½ inches [3.5 to 4 cm]). Thread-legged bugs feed on a diversity of

Photo: D. Shetlar

△ Here is a thread-legged bug in the genus Barce *extending with its pair of raptorial front legs ready to strike.*

Photo: Jon M. Yuschock

△ *Thread-legged bugs attack many pests, but some also feed on other natural enemies. This* Emesaya brevipennis *is able to hunt within a spider's web without becoming entangled.*

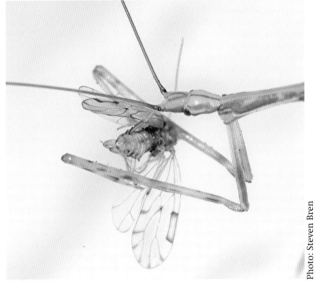

Photo: Steven Bren

△ Here is a thread-legged bug in the genus Pseudometapterus *feeding.*

Photo: Richard Hoyer (louisvillepics.com)

△ *Jagged ambush bugs in the genus* Phymata *often hunt in flowers and feed on both pest and other beneficial arthropods.*

pest arthropods and on some beneficial arthropods. For example, some are able to hunt within spider webs, attacking both the spider and potentially its catch.

Ambush Bugs

Ambush bugs are more stout-bodied than assassin bugs or thread-legged bugs. They are often brightly colored with front legs that are more enlarged than thread-legged or assassin bugs and range in size from 7 to 18 mm. The nymphs of ambush bugs look similar to the adults but without fully formed wings. Species within three genera, *Lophoscutus, Phymata,* and *Macrocephalus,* are found in the United States. Ambush bugs are often found hunting in flowers, and, as they are generalist predators, they will feed on pollinators as well as pests.

Assassin Bugs

Assassin bugs are typically dark brown, black, or gray, although some, particularly in the genus *Zelus,* are brightly colored. They have long, narrow heads and a diamond-shaped body. The nymphs of assassin bugs look similar to adults, but without fully formed wings. Some assassin bugs found in the garden include wheel bugs, spined or spiny assassin bugs, masked hunters, the *Zelus* species, and corsairs.

WHEEL BUGS (*ARILUS*)

Wheel bugs are one of the largest true bugs in North America, measuring up to 1½ inches (4 cm) in length. They are gray to black and take their name from a characteristic half-wheel of spines that protrude from the top of their thorax. Although they are fairly common,

Photo: Ilona L.

△ *This* Sinea diadema *shows the diamond-shaped body common to many species within the family Reduviidae.*

▽ *Adult wheel bugs in the genus* Arilus *are one of the largest true bugs in North America at up to 1½" (4 cm) in length.*

Photo: Mary M. Gardiner

Photo: Odin Toness

△ *The milkweed assassin bug (Zelus longipes) has a slender, bright orange body with dark markings and dark legs and antennae.*

Photo: David Gottlieb

△ *Here a nymph of the pale green assassin bug (Zelus luridus) feeds on a leafhopper.*

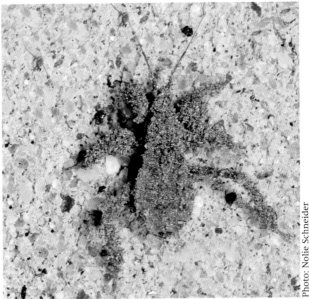

Photo: Nolie Schneider

△ *The name masked hunter, given to the nonnative Reduvius personatus, comes from the nymphs' habit of coating themselves in dirt and debris. This provides camouflage and protects the young reduviids from potential predators.*

Photo: Robin McLeod (whatonearthphotos.com)

△ *Adult masked hunters (Reduvius personatus) are sometimes found indoors. They inflict a painful bite, so handle with care! This individual has captured an earwig (Dermaptera).*

wheel bugs are rarely seen in the garden due to their camouflage coloration and sit-and-wait hunting strategy. Spined and spiny assassin bugs (*Sinea diadema* and *Sinea spinipes*) are light gray to brown and smaller than wheel bugs, measuring about ¾ of an inch (2 cm) in length. Assassin bugs in the genus *Zelus*, such as the milkweed assassin bug (*Zelus longipes*), leafhopper assassin bug (*Zelus renardii*), pale green assassin bug (*Zelus luridus*), and four-spurred assassin bug (*Zelus tetracanthus*), have slender bodies and can be brightly colored with green, orange, or yellow bodies and dark markings on their wings, thorax, and legs. The masked hunter (*Reduvius personatus*) is an introduced species that is common in the eastern and central United States. They are named for the behavior of the immature nymph insects, which cover their bodies in soil and other debris to camouflage and protect themselves from predators.

The adults are dark brown to black. You can sometimes find a masked hunter in your home. They are known to feed on bed bugs (Cimicidae: *Cimex lectularius*) and bat bugs (Cimicidae: *Cimex adjunctus*).

CORSAIRS (*MELANOLESTES*)

Corsairs, which measure ¾ to 1 inch (2 to 2.5 cm) in length, are unique among the Reduviidae in that males and females look and behave differently. Males have fully formed wings and can disperse readily to hunt and search for mates, whereas females lack wings or have small nonfunctional wings. Females are found hunting for ground-dwelling prey in gardens on the soil surface or within mulch or leaf litter.

△ *Male corsairs* (Melanolestes) *are dark in color with fully functional wings.*

△ *Female corsairs* (Melanolestes) *lack wings, or, as in the case of this individual, have small, nonfunctional wings.*

True Bug Predators in Garden Water Features

There are several families of predatory Hemiptera that may colonize garden ponds and other water features and provide biological control of the pests found within them, including mosquito larvae (Culicidae) and pests of aquatic pond plants. Like terrestrial Hemiptera, these aquatic predators feed on prey both as adults and as nymphs. Nymphs will molt several times to reach the adult stage, and they will overwinter as either late instar nymphs or adults. One or more generations are possible within a single growing season.

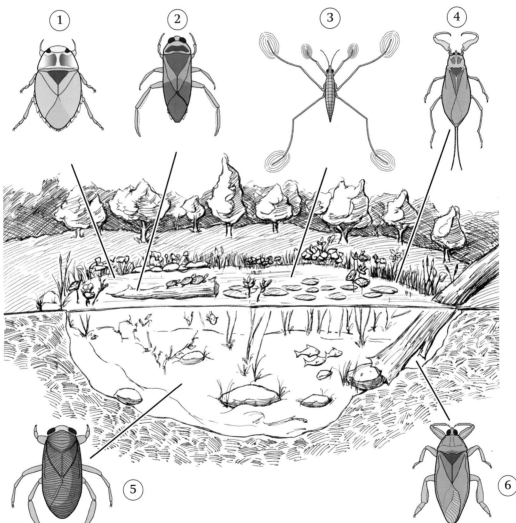

△ 1. *Creeping water bugs (Naucoridae) can be found along the margins of ponds.* 2. *Back swimmers (Notonectidae) swim on their backs at the pond surface.*
3. *Water striders (Gerridae) can skate along with water's surface.* 4. *Water scorpions (Nepidae) are typically found foraging in the vegetation on pond margins.*
5. *Water boatmen (Corixidae) hunt in the sediments at the bottom of ponds.* 6. *Giant water bugs (Belostomatidae) can be found below the surface of the water. They also come to lights at night.*

△ *The giant water bugs (Belostomatidae) are also called toe biters because they are found below the surface of ponds and are voracious predators.*

Giant Water Bugs (Belostomatidae)

Giant water bugs are large (¾ to 1¾ inches [2 to 4.5 cm]), oval shaped, and flat, with enlarged raptorial front legs and hind legs lined with swimming hairs. Females lay eggs either on emergent vegetation or on the backs of males. The nymphs undergo several molts before reaching the adult stage and overwinter as adults in mud. Belostomatidae live below the surface of ponds and are voracious predators that attack prey many times their size, including fish and frogs, but the majority of their diet consists of insect larvae. Some species of these true bug predators are also called toe biters because they produce hydrolytic enzymes that they inject into their prey, which causes pain and swelling if you inadvertently step on one.

Photo: Stephen Luk

△ *Water boatmen are oval shaped with short front legs, which they use to disturb the bottom of ponds and streams in search of prey.*

Photo: D. Shetlar

△ *Water striders (Gerridae) take their name from their ability to skate on the surface of the water. As adults, striders may or may not have wings.*

Water Boatman (Corixidae)

Water boatman are small (3 to 11 mm) with an oval-shaped body and short front legs modified to disturb sediments on the bottoms of ponds and streams to search for prey. They have oarlike hind legs and swim dorsal-side up. They breathe using an air bubble they hold under their wings and must break the surface of the water to renew their air supply. They lay eggs on submerged debris and sometimes even on other organisms such as crayfish. Water boatmen are strong fliers and can easily disperse among ponds and other bodies of water.

Water Striders (Gerridae)

Water striders have a long, slender body (3 to 18 mm) and long legs. They take their name from the way they skate along the surface of the water, using their middle and hind legs. They can also dive underwater and store air in hairs on their body. Water striders disturb the water's surface to detect possible prey and use their front legs to capture insects that either inhabit the water's surface or those that become trapped there. Gerridae may be winged or wingless as adults.

△ *Creeping water bugs (Naucoridae) are found on the margins of ponds and in marshy areas. They are flat and oval shaped with enlarged front legs for grasping prey.*

Creeping Water Bugs (Naucoridae)

Creeping water bugs are small (6 to 15 mm) and typically dwell within the margins of ponds and marshy areas, where they can be found along the water's edge or on emergent vegetation in search of aquatic insects and mollusks to feed on. They are flattened and oval in shape with enlarged raptorial front legs. Females attach their eggs to submerged pebbles, vegetation, or debris.

Water Scorpions (Nepidae)

Water scorpions are elongated predators that range in size from 5 to 55 mm and can be found within the vegetation that thrives along pond margins. Their leaf or sticklike bodies offer very good camouflage at the edges of ponds. Water scorpions are sit-and-wait predators that use their mid and hind legs to attach to vegetation and hang with their head and raptorial front legs underwater to hunt.

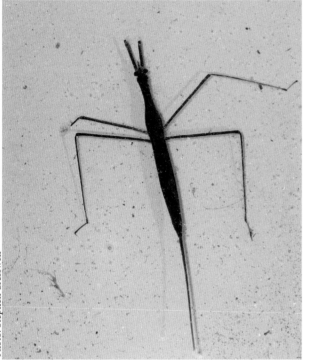

Photo: Stephen Cresswell

△ *Water scorpions (Nepidae) use an appendage extending from their abdomen to breathe while underwater. Their leaf- or sticklike bodies offer good camouflage from predators.*

Photo: Kim Moore (kimssight.zenfolio.com)

△ *Backswimmers look similar to water boatmen but are found swimming on their backs, using their long, oarlike hind legs for propulsion.*

While underwater, they breathe using an appendage that extends from their abdomen. Adults disperse among water bodies by flying, and females lay eggs either on floating vegetation or in the mud along the water's edge.

Backswimmers (Notonectidae)

Backswimmers are similar in size (5 to 15 mm) and body shape to water boatman, with large eyes, an oval-shaped body, and long oarlike hind legs. But as their name implies, backswimmers swim on their backs. Notonectidae swim at or near the surface of the water and store air on the ventral side of their abdomen as well as under their wings. They prey on a diversity of aquatic arthropods and small vertebrates. Backswimmers lay their eggs on aquatic plants or submerged debris. Backswimmers are strong fliers.

6

Lacewings and Other Net-Winged Predators (Neuroptera)

The order Neuroptera includes some of the strangest-looking garden predators, but a closer look reveals both their beauty and ferociousness. All adult insects in this order have four wings that are either the same size or nearly so with intricate veins, giving them a lacy appearance. When not in flight, Neuroptera often carry their wings tentlike over their abdomen. They have slender antennae, chewing mouthparts, and large eyes. Many adults are active at dawn and dusk or during the night and are attracted to lights. Neuroptera develop by complete metamorphosis with egg, larval, pupal, and adult stages. Some species are predators as both adults and larvae, while others are predators only during the larval stage. There are several families of Neuroptera that can be found in home gardens. The brown and green lacewings (Hemerobiidae and Chrysopidae) are most common. Others, such as owlflies (Ascalaphidae) and mantidflies (Mantispidae), are rarer.

Owlflies (Ascalaphidae)

Adult owlflies are large insects, measuring about 1 to 2 inches (3 to 5 cm) in length, that resemble dragonflies. They are more common in the southern United States, but do occur in the northern states as well. Owlflies have large eyes, long, robust abdomens, and long, net-veined wings. They have long antennae with a ball or club shape at the tip. Adults are active at dawn and dusk as well as during the night. When at rest, they press their bodies and wings against branches or stems. Some species, such as *Ascaloptynx*

appendiculata, also rest with their abdomen held outward to look like a twig or branch. Adult females lay clusters of eggs in spirals or rows. They feed on a large number of flying insects and catch their prey on the wing.

Owlfly larvae are ambush predators. Some resemble leaves or twigs and forage on vegetation. Other species forage in leaf litter and sometimes cover their bodies with soil or plant debris. These larvae wait with their jaws open for unsuspecting prey. Owlfly larvae pupate in the leaf litter within a round, silk cocoon.

Photo: Giff Beaton

△ *Owlflies (Ascalaphidae) have long antennae with a ball or club shape at the tip. While at rest, some hold their long abdomen outward from their perch.*

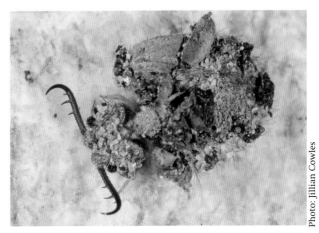

Photo: Jillian Cowles

△ *All you can see are jaws! This* Ululodes arizonensis *larva has achieved great camouflage by covering its body with soil and debris. The larva will sit and wait for passing prey to come within its grasp.*

Green Lacewings (Chrysopidae)

The green lacewings are a large family with species found throughout the United States. Depending on the species, adult green lacewings may be predators or pollen feeders, or they may feed on honeydew, which is a sugar-rich substance secreted by aphid (Aphididae) pests. All lacewing larvae are predatory.

The adults range from 8 to 25 mm in length and, like the *Chrysopa* and *Chrysoperla* – two common genera within agricultural and garden habitats – are typically bright green. A few genera, such as *Eremochrysa,* are reddish or brown. Green lacewings have four clear, intricately veined wings, a long slender abdomen, long antennae,

and large iridescent eyes. They have excellent hearing; organs at the bases of their wings can even detect ultrasound calls from bats. If a bat comes near, the lacewing will close its wings and drop to the ground to avoid the potential predator. Although the more than eighty species of green lacewings found in the United States may look very similar to us, the lacewings themselves are easily able to locate mates of their own species through the courtship songs they produce by making vibrations with their bodies. Another common name for green lacewings is stink flies due to the ability of species in the genus *Chrysopa* to release a bad odor when disturbed. This compound is also used in courtship, so perhaps the lacewings do not find it as foul.

After mating, females deposit single stalked eggs or small groupings of stalked eggs on vegetation. It is believed that green lacewings lay eggs in this manner to protect the eggs from parasitism and predation and to prevent the first siblings that hatch from cannibalizing the larva that hatch later. When the larvae emerge from their egg, they crawl down the stalk and disperse in search of other prey, without consuming their brothers and sisters.

Green lacewing larvae are voracious aphid predators and are sometimes called aphid lions. They are slender, mottled brown and cream, and have large sickle-shaped jaws. Different species can look very similar as larvae with

Photo: D. Shetlar

△ *Chrysopidae adults are most often green with four intricately veined wings, a slender abdomen, long antennae, and large iridescent eyes.*

one main distinction: some are "naked" and others are "debris-carrying." The debris-carrying larvae cover themselves with debris such as plant parts, waxy secretions, or the remains of prey. The debris acts as camouflage from predators. Both naked and debris-carrying larvae have sickle-shaped grooved jaws. To eat, the larva will pierce an insect with its jaws and lift the prey into the air, allowing its liquid contents to flow down the grooves and into its mouth. When ready to pupate, the larvae spin a silken cocoon around plant leaves or stems.

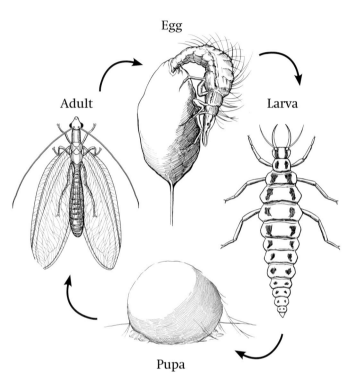

Egg

Adult

Larva

Pupa

▷ Female green lacewings (Chrysopidae) lay stalked eggs singly or in clusters on plant stems and leaves. When a larva emerges from its egg, it climbs down the stalk and disperses in search of prey. The eggs being stalked prevent an individual larva from consuming its brothers and sisters still within their eggs.

Photo: Mary M. Gardiner

△ Green lacewing larvae are sometimes called aphid lions because they are voracious predators with sickle-shaped jaws.

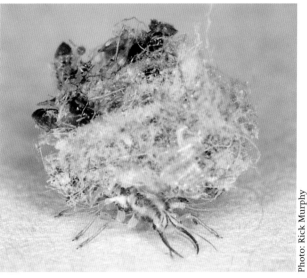

Photo: Rick Murphy

△ The larvae of some green lacewing species cover themselves with debris, including plant parts, wax, or the remains of their prey to camouflage the larva from potential predators.

Dustywings
(Coniopterygidae)

Brown Lacewings
(Hemerobiidae)

Dustywings adults are very easy predators to miss in the garden because they are minute to small in size (1 to 4 mm) and look like small moths or whiteflies. To determine the difference between these look-alikes, watch carefully to see how they hold their wings; Dustywings hold them tentlike over their abdomen. Their bodies and translucent wings have a white or light gray appearance due to a covering of waxy scales that look like a light layer of dust. Adults are most often found on trees and shrubs. They are active at dawn and dusk and are attracted to lights. Females will lay eggs one at a time on bark or leaves near pest infestations. The larvae are slender with light and dark markings. They have two straight tubular mouthparts that are used to pierce prey and ingest its liquid contents. Dustywings adults and larvae are predators of small soft-bodied prey such as insect eggs, whiteflies (Aleyrodidae), aphids (Aphididae), and spider mites (Tetranychidae).

Brown lacewings look very similar to green lacewings (Chrysopidae) except they are almost always brown to gray and typically smaller, measuring 2 to 13 mm in length. Adults have four intricately veined wings of equal size, a slender body, large eyes, and long, thin antennae. Unlike green lacewings, the eggs of brown lacewings are not stalked; instead, females deposit them directly on plant material. Some species are able to lay hundreds of eggs. Adults are active at dusk and after dark, when temperatures are cooler than many other natural enemies will tolerate. Brown lacewing adults are predatory. Those in the genera *Hemerobius* and *Micromus* feed on aphids (Aphididae) and other soft-bodied prey and are commonly found within

Photo: Shu-Shin Chin

△ *Dustywings (Coniopterygidae) are covered in white, waxy scales that give them a dusty appearance.*

Photo: Lynette Schimming

△ *Brown lacewings (Hemerobiidae), such as this individual in the genus* Hemerobius, *have four intricately veined wings of equal size, a slender brown body, large eyes, and long antennae.*

Mantidflies (Mantispidae)

home landscapes, particularly those that are wooded or near wooded areas. Brown lacewings in the genus *Sympherobius* look similar but are smaller, measuring 2 to 6 mm, and feed on scale insects (Coccoidea). You might find brown lacewing adults on a porch at night because they are attracted to lights.

Brown lacewing larvae look very similar to naked green lacewing larvae; they are brown, white, and gray with a long, slender body and sickle-shaped mandibles. When ready to pupate, the larvae spin a loose cocoon, attaching it to plant material.

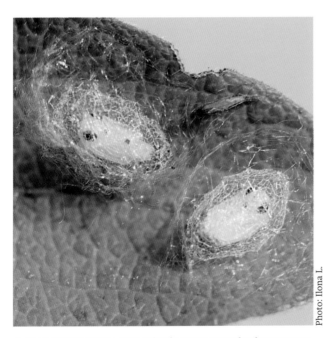

Photo: Ilona L.

△ *Brown lacewings larvae construct a loosely woven pupa on plant leaves or stems.*

Mantidflies are a rare garden predator, but they had to be included here because they are just so bizarre looking! They range in size from small to large (5 to 47 mm) and are named for their resemblance to mantids (see chapter 4). Like mantids, mantidflies have enlarged raptorial front legs that they use to grasp prey. They also have a long, slender thorax, triangular-shaped head, and large eyes. The adults are active day and night and are attracted to lights. Most mantidflies have clear net-veined wings with bodies that range from green (green mantidfly, *Zeugomantispa minuta*) to yellow and brown (*Dicromantispa interrupta*). The wasp mantidfly (*Climaciella brunnea*) resembles both a mantid and a paper wasp (*Polistes*, see chapter 9). This species has a brown body with yellow markings and a leathery brown portion of their front wing. Mantidflies can be found on flowers because they feed on pollen and nectar as adults, but they also feed on a large number of different arthropods, including both pests and beneficial species.

Females lay clusters of several hundred eggs, most likely because their larvae have a difficult path to adulthood. These light beige to brown segmented larvae look like grubs, but they are actually parasitoids of spiders that develop by feeding within a spider egg sac. The larva may actively search for a spider egg sac on its own, in which case, both active hunting and web-building spider eggs may be parasitized. In other cases, the larva will wait for a hunting spider to come within reach and hitch a ride. When the spider begins constructing an egg sac, the larva will enter it. If the mantidfly did not initially select a female, the larva is even able to move between spiders while the spiders are mating.

△ *Mantidflies (Mantispidae) are so jumbled in appearance that they look as if a student, desperate to turn in an insect collection for a class project, may have created them by gluing a bunch of spare insect parts together! This family includes the green mantidfly (1) (Zeugomantispa minuta), yellow and black mantidfly (2) (Dicromantispa interrupta), and the wasp mantidfly (3) (Climaciella brunnea).*

△ *Mantidfly females deposit hundreds of eggs in clusters.*

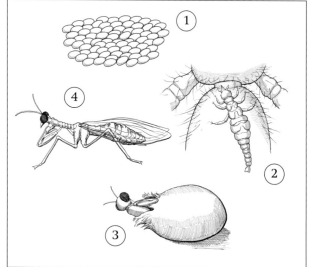

△ *1. Mantidfly females lay several hundred eggs to ensure that a few of their offspring make it to the adult stage. 2. The hatching larvae must first find a spider, which they will ride until until the spider lays eggs. If the spider is a male, the larva will transfer onto a female during mating. 3. When the female spider lays her eggs, the mantidfly larva burrows inside and completes its larval stage by consuming them. 4. The mantidfly pupates in the egg sac of the spider and emerges as an adult.*

Antlions or Doodle Bugs (Myrmeleontidae)

Myrmeleontidae is the largest family within the order Neuroptera. As the name suggests, antlion larvae do feed on ants, but these predators consume a large number of other arthropods as well. Adult antlions have long, slender bodies, reaching 1½ to 3 inches (4 to 8 cm) in length. They have long net-veined wings and short, curved antennae. In some species, the wings are entirely clear; on others such as *Glenurus* and the spotted-winged antlion, *Dendroleon obsoletus*, the wings have black and white patterning. Adults are active at night and are attracted to lights. Females lay eggs in the soil or on the soil's surface.

Antlion larvae have an oval-shaped gray to brown body and large jaws. The most famous larvae in the family Myrmeleontidae are those that build pits in sand to catch their prey. The larva constructs a funnel-like pit and settles in at the bottom. When an insect falls into its trap, the antlion larva will throw sand onto its victim, causing it to tumble deeper toward the base of the pit. The larva will grasp the insect with its large jaws and pull it below ground to consume it. Not all antlion larvae hunt using a pit. Some actively seek prey in the leaf litter while others wait in protected sites such as crevices in tree bark to ambush their prey. They pupate in a belowground cocoon constructed of sand and silk.

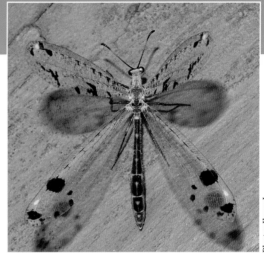

Photo: Ilona L.

△ Antlion adults (Myrmeleontidae) have slender bodies, long wings, and short, curved antennae. Some have clear wings, while others, such as the spotted-winged antlion Dendroleon obsoletus, *have patterned wings.*

Photo: D. Shetlar

△ Antlion larvae are covered with bristles and have large sickle-shaped mandibles.

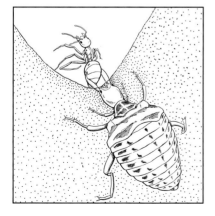

◁ Some antlion larvae use a trap to hunt for insect prey. These larvae construct a conical pit in the sand and then nestle themselves below ground at the base of the pit. When an unsuspecting insect falls into the trap, the larva will toss sand at it, causing the insect to fall deeper into the pit and become trapped. The antlion will then grasp the insect with its large jaws and pull it below ground to consume it.

Photo: D. Shetlar

△ Here is a group of several pits dug by antlion larvae.

7

Beetles (Coleoptera)

When in doubt with an insect identification, guess beetle! After all, beetles represent 40 percent of described insect fauna. These easily adaptable creatures occupy nearly every habitat on earth and can be found in trees and on herbaceous plants, on the soil surface, below ground, and in water features around the yard. Beetles develop through complete metamorphosis and have an egg, larval, pupal, and adult stage.

If you have inadvertently – or perhaps purposefully – stepped on a beetle, the crunching sound that followed is evidence of just how hard a beetle's body is. The name Coleoptera roughly translates to sheath wing. All beetles share a common wing characteristic – hardened front wings called elytra that cover and protect the membranous hind wings that are actually used for flying.

Certainly many beetles are pests, and as a group, they have been known to consume any available plant part, including roots, leaves, stems, flowers, fruits, and seeds. Thus, a gardener's fondness for the beetles can wane after a day of removing Colorado potato beetle eggs and larva from rows of plants or pulling out squash plants killed by bacterial wilt, a virus transmitted by cucumber beetles. Yet, this order does include a vast diversity of beneficial species, including many that attack garden pests. Predatory beetles have chewing mouthparts and feed on prey as both a larva and adult, but some species consume pests only during the larval stages.

The wide array of beetle species can be overwhelming. For the purposes of this book, we will focus on those most likely to be found foraging within home landscapes, including tiger and ground beetles (Carabidae) and lady beetles (Coccinellidae), two large groups of important garden predators, as well as other unique beetles.

Soldier Beetles (Cantharidae)

Soldier beetles (1 to 15 mm) look similar to fireflies (Lampyridae), but with a head not covered completely by their pronotum when viewed from above. Also, unlike fireflies, Cantharidae are unable to produce light. The adults have long, thin antennae and leathery wings that cover a long, slender abdomen. Species in several soldier beetle genera such as *Atalantycha* (6 to 9 mm), *Cantharis* (½ to ¾ of an inch [1 to 2 cm]), *Podabrus* (½ to ¾ of an inch [1 to 2 cm]), and *Rhagonycha* (5 to 7 mm) have dark brown or black elytra with bright colors such as orange, yellow, or red present mainly on their pronotum, head, and abdomen and/or edges of the elytra.

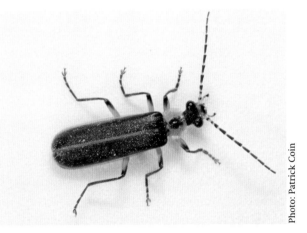

△ Many beetles in the family Cantharidae are slender with dark, leathery wings and long antennae. Like this Podabrus *species, they often have bright colors present on their pronotum, a plate that covers their thorax.*

In the genus *Chauliognathus*, species such as the goldenrod soldier beetle (*Chauliognathus pennsylvanicus*), Colorado soldier beetle (*Chauliognathus basalis*), and margined leatherwing (*Chauliognathus marginatus*) are very attractive with larger amounts of color present on their elytra.

These bright colors warn potential predators to stay away. Soldier beetles also produce cantharidin, a defensive chemical that acts as a blistering agent. The beetles emit cantharidin when they feel threatened by other arthropods and birds. Adult beetles are active during the day and are commonly found on plants, especially in flowers where they feed on pollen and nectar. Many species—but not all—are insect predators as adults.

Soldier beetle larvae live in leaf litter and prey upon many soil-dwelling arthropods and insect eggs. The beetles have flattened, elongated bodies and are relatively uniform in width. They have short legs and are covered with hairs, which give them a velvety appearance.

△ The goldenrod solider beetle (*Chauliognathus pennsylvanicus*) *blends in well with the goldenrod flowers where it is often seen feeding and mating.*

△ *Cantharidae larvae are predators of soil-dwelling insects. They have an elongated body that appears velvety.*

Ground Beetles and Tiger Beetles (Carabidae)

The family Carabidae is very large and includes two groups – ground beetles and tiger beetles. Ground beetles vary in size from small to large (1 to 35 mm) and are often black or metallic and shiny with indented ridges or dots on their elytra. Ground beetles may also be light brown with or without dark brown spots. Adults have a wedge-shaped head and thorax that allows them to burrow within garden mulch, leaf litter, or soil. They have large eyes, long, thin antennae, and large sickle-shaped mandibles. Ground beetles have a diverse diet, feeding on arthropods, millipedes, snails, slugs, and plant seeds. Given their name, it's not surprising that most adult ground beetles are fast runners that actively hunt for their prey along the soil surface or within leaf litter or mulch, but some do climb vegetation in search of prey. Ground beetles typically hunt after dark. During the day, you may uncover them when weeding or by looking under stones or bark mulch in the garden.

Ground beetle larvae have large heads, large jaws, and heavily sclerotized abdomen that are covered with plates and taper toward the end. Most have walking legs and actively hunt prey or seeds, but some are parasitoids of other arthropods and appear more grublike.

Photo: Patrick Coin

△ *Most Carabidae larvae are dark with an elongated body covered with dark plates. They have large mandibles. Some are parasitoids and look more like light-colored grubs. This particular larva is feeding on a snail.*

Several ground beetles are common within agricultural habitats, including home gardens and residential landscapes. *Pterostichus* and *Poecilus* genera are among the most common and important garden predators. The genus *Pterostichus* has 150 species found in the United States. *Pterostichus melanarius* is a common species native to Europe that was introduced in the United States. Measuring ½ to ¾ of an inch (1 to 2 cm), *Pterostichus melanarius* (**1**) is black and feeds on many types of prey including caterpillars, beetle grubs, fly maggots, aphids (Aphididae), earthworms, slugs, and snails.

Poecilus chalcites (**2**) (½ to ¾ of an inch [1.5 to 2 cm]) feeds on many crop pests, including rootworms (Chrysomelidae), cutworms (Noctuidae), and armyworms (Noctuidae), and is active in the early spring, which makes it an important biological control agent.

The bigheaded ground beetle, *Scarites subterraneus* (**3**), is also common within cultivated soils. This large (½ to ¾ of an inch [1.5 to 2 cm]) shiny black ground beetle has a flattened, elongated body, large head, and front legs designed for digging.

Gardeners will find these beetles under rocks or in bark mulch where they build burrows and feed on many types of insect prey including cutworms (Noctuidae) and fly larvae.

The four-spotted ground beetle (*Bembidion quadrimaculatum*) (**4**), which measures 2 to 4 mm in length, is among the smallest ground beetles found in gardens. With a black body, light legs, and black elytra with four light-colored spots, this helpful beetle feeds on the eggs of many pests, including the seedcorn maggot (Anthomyiidae: *Delia platura*) and cabbage maggot (Anthomyiidae: *Delia radicum*).

The large foliage ground beetle, *Lebia grandis* (**5**) (8 to 13 mm), is an example of a Carabidae species that is active in the plant canopy as an adult. As a predator of the Colorado potato beetle (Chrysomelidae: *Leptinotarsa decemlineata*), a serious potato pest, and the false potato beetle (Chrysomelidae: *Leptinotarsa juncta*), which can wreak havoc on eggplant, this beetle is a gardener's best friend. Adult *Lebia grandis* feed on larvae and eggs they find on plant leaves and stems. The larvae are parasitoids of pupae. When they hatch from their egg, they locate a beetle pupa in the soil and begin feeding, eventually killing it. The larva will then pupate and emerge from the soil as an adult.

Snail-eating ground beetles, *Scaphinotus* (**6**), are medium to large (½ to 1¼ inches [1.5 to 3 cm]) Carabidae that specialize on snails and slugs. They have slender heads with adapted mandibles and thorax that enable them to eat snails from within their shell.

Some of the largest ground beetles in the United States are in the genus *Calosoma*. These beetles can be found on shrubs and trees within residential landscapes. The forest caterpillar searcher (*Calosoma sycophanta*) (**7**) was introduced from Europe to control the gypsy moth (Erebidae: *Lymantria dispar*), a significant pest of trees throughout forested areas of the northern United States. This large beetle measures ¾ to 1¼ inches (2 to 3 cm) and has bright green elytra and an iridescent blue to black thorax. They live for several years, and both larvae and adults climb trees in search of gypsy moths and other forest caterpillars.

The fiery searcher *Calosoma scrutator* (**8**) measures ¾ to 1¼ inches (2 to 3 cm), is bright green as an adult, and feeds on tent caterpillars (Lasiocampidae) and other caterpillar pests of shrubs and trees.

The fiery hunter *Callisthenes calidus* (**9**) has a preference for caterpillars, specifically armyworms (Noctuidae) and cutworms (Noctuidae). Measuring ¾ to 1 inch (2 to 2.5 cm), it is black with rows of indented golden or pink spots.

The European ground beetle, *Carabus nemoralis* (**10**) (¾ to 1 inch [2 to 2.5 cm]), is an introduced species commonly found foraging in cultivated soils. Beetles in this genus have been reported to exhibit courtship behavior, which is very rare among beetles. Some ground beetles, including a few *Carabus* species, are also known to hunt on the margins of ponds and other water bodies for snails and insect larvae. They can store air beneath their wing covers to replenish their supply when submerged. Larvae of these species are also semi-aquatic and can hunt with their heads underwater.

Many of the ground beetles in the genus *Harpalus* feed on both arthropod prey and seeds. The murky ground beetle *Harpalus caliginosus* (**11**) is a very common, large (¾ to 1 inch [2 to 2.5 cm]) beetle that is black or reddish black. The larvae feed on the Colorado potato beetle larvae (Chrysomelidae: *Leptinotarsa decemlineata*) as well as plant seeds, while adults feed on a variety of insect prey and can sometimes become a pest of strawberries. The strawberry seed beetle *Harpalus rufipes* (½ to ¾ of an inch [1 to 2 cm]) is also common in gardens. This black beetle with light orange legs and yellow hairs on the base of its elytra feeds on weed seeds but can also consume the seeds off the outside of strawberry fruits. The Pennsylvania ground beetle *Harpalus pensylvanicus* (**12**) is found throughout the United States and feeds on caterpillars and seeds. The genus *Amara* (5 to 11 mm) (**13**) predominately eats seeds as well. Many species are common in gardens and are known to feed on several weed seeds.

Bombardier beetles (*Brachinus*) (**14**) are medium-size (5 to 13 mm) with bluish metallic elytra and orange legs, thorax, head, and antennae. Often found near the edges of streams or ponds, adults hunt at night and feed on a diversity of arthropod prey. Females lay single eggs within mud cells that they attach to rocks or plants and the larvae are parasitoids of aquatic beetles. These beetles are named for their ability to spray would-be attackers with hydroquinone and hydrogen peroxide! These compounds are stored in separate organs within the beetle, and when combined with water and enzymes, the mixture exceeds the boiling point of water and is ejected out the back of the beetle. A blast of this spray is often deadly to arthropod predators and can cause pain and skin irritation for humans.

Getting to Know the Ground Beetles

1

Photo: Stephen Luk

Common Ground Beetle
(Pterostichus melanarius)
½ to ¾ of an inch (1.5 to 2 cm)

2

Photo: Roxanne S. Bernard

Common Ground Beetle
(Poecilus chalcites)
½ to ¾ of an inch (1.5 to 2 cm)

3

Photo: Ryan Kaldari

Bigheaded Ground Beetle
(Scarites subterraneus)
(1.2 to 1.4 mm)

4

Photo: Yves Dubuc

Four-Spotted Ground Beetle
(Bembidion quadrimaculatum)
(2 to 4 mm)

5

Photo: John and Jane Balaban

Large Foliage Ground Beetle
(Lebia grandis)
(8 to 13 mm)

6

Photo: Gary McDonald

Snail-Eating Ground Beetle
(Scaphinotus)
½ to 1¼ inches (1.5 to 3 cm)

7

Photo: Kevin Schick

Forest Caterpillar Hunter
(Calosoma sycophanta)
¾ to 1¼ inches (2 to 3 cm)

8

Photo: Jason D. Roberts

Fiery Searcher
(Calosoma scrutator)
1 to 1½ inches (2.5 to 4 cm)

9

Photo: Brandon Woo

Fiery Hunter
(Callisthenes calidus)
¾ to 1 inch (2 to 2.5 cm)

10

Photo: John and Jane Balaban

European Ground Beetle
(Carabus nemoralis)
¾ to 1 inch (2 to 2.5 cm)

11

Photo: Patrick Coin

Murky Ground Beetle
(Harpalus caliginosus)
¾ to 1 inch (2 to 2.5 cm)

12

Photo: Charley Eiseman

Pennsylvania Ground Beetle
(Harpalus pensylvanicus)
¾ to 1 inch (2 to 2.5 cm)

13

Photo: Ryan Kaldari

Sun Beetle
(Amara)
(5 to 11 mm)

14

Photo: Lon Brehmer and
Enriqueta Flores-Guevara

Bombardier Beetle
(Brachinus)
(5 to 12 mm)

Tiger Beetles

Tiger Beetles (subfamily Cicindelinae) look quite a bit different from other beetles in the family Carabidae. They are often very colorful beetles with slender metallic bodies and long legs. Tiger beetles can be brown or black, such as night-stalking tiger beetles (*Omus*, ½ to ¾ of an inch [1.5 to 2 cm]) and giant tiger beetles (*Amblycheila*, ¾ to 1½ inches [2 to 4 cm]) or bright metallic green, purple, blue, pink, red, or golden such as the metallic tiger beetles (*Tetracha*, ¾ to 1 inch [2 to 2.5 cm]) and common tiger beetles (*Cicindela*, ½ to ¾ of an inch [1 to 2 cm]).

All tiger beetles have large eyes, long antennae, and large sickle-shaped mandibles. Tiger beetles are active hunters that can run so fast along the soil surface that they sometimes lose track of the prey they are hunting and need to stop momentarily to search for it. They can fly, but most species only move short distances on the wing. Except for night-stalking tiger beetles, most tiger beetles are active during the day. Some species are attracted to lights at night. As adults, these beetles construct burrows where they spend their days and nights. Most eat live prey, feeding on a number of different arthropods. Some will scavenge for dead prey as well.

Tiger beetles are typically found in sandy soil or forested areas. Some species are active in mid-summer, while others are active in the spring and fall. When ready to mate, males will sprint quickly toward the female, jump onto her back, and hold her by fitting his mandibles into unique grooves on the sides of her thorax. If the female does not reject the advance, they will mate, and the male will continue to hold onto the female while she moves around in the environment before laying her eggs. This behavior prevents other males from fertilizing her eggs. Females lay eggs one-by-one in the soil, covering them up to help reduce egg predation. Because the larvae will construct burrows at these sites, females will only lay eggs in sites with the appropriate temperature, soil, and moisture conditions.

The larvae of tiger beetles are very unique. They have grublike bodies and a heavily sclerotized head with large mandibles. They construct a narrow burrow that runs perpendicular to the ground. The larva can anchor itself in the burrow using hooks that extend from a humplike projection on the dorsal side of its abdomen. They have six eyes and are able to discern prey coming toward their burrow. The larva waits in its burrow with its head at the soil surface, and when prey comes within striking distance, extends out of the burrow to grasp potential prey and pull it into the burrow where all ingestible parts will be consumed and anything remaining will be tossed out.

Most pits are round and even with the soil surface, but some larvae construct a conical pit around their burrow or add sticks, pine needles, or other inviting landing substrates for potential prey. Larvae go through three molts before they construct a pupal chamber belowground near their burrow and emerge as adults. It can take up to three years for a tiger beetle larva to mature to the adult stage.

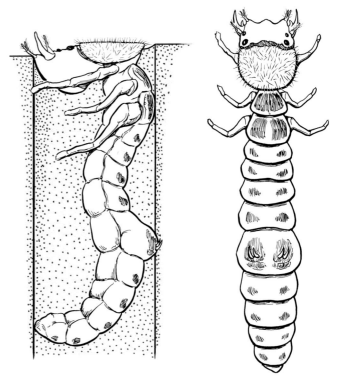

▷ *Tiger beetle larva hunt from a burrow constructed perpendicular to the ground. They have a humpbacked body that helps to anchor the larva within its burrow. Their flat head is held even with the soil surface until they strike out and grasp prey with their large mandibles. Once the prey is in their grasp, they retract into their burrow to consume it. The burrow of a tiger beetle larva is often round and even with the soil surface, but some species add perches for potential prey, sticks, needles, or other materials to surround the entrance. Others construct a small pit surrounding their burrow to cause passing insects to fall toward the hungry larva. Tiger beetle larvae can remain in their burrow for up to three years before pupating and emerging as an adult.*

The Beauty of Tiger Beetles

Giant Tiger Beetle
(Amblycheila)
¾ to 1½ inches (2 to 4 cm)

Common Tiger Beetle
(Cicindela)
½ to ¾ of an inch (1.5 to 2 cm)

Night-Stalking Tiger Beetle
(Omus)
½ to ¾ of an inch (1.5 to 2 cm)

Metallic Tiger Beetle
(Tetracha)
¾ to 1 inch (2 to 2.5 cm)

Lady Beetles (Coccinellidae)

Lady beetles, also called ladybugs or lady bird beetles, are one of the most beloved garden predators. This large family includes many species that feed on insect pests such as aphids (Aphididae), spider mites (Tetranychidae), thrips (Thripidae), and scales (Coccoidea) in home landscapes. Lady beetle adults are oval to round in shape with a head that is almost completely concealed from above by the pronotum, which is a plate that covers their thorax. Although they are typically thought of as having red elytra with black spots, lady beetle species vary widely in color and may or may not have spots. Lady beetle larvae are also predatory and can be found foraging in the same habitats as adults.

The larvae of the majority of species are elongated and slender with short legs and dark spiny bodies with white, yellow, or orange markings. The larvae of the spider mite destroyer (*Stethorus punctum*) are completely black and covered with long hairs. The larva of the twice-stabbed lady beetle (*Chilocorus stigma*) and metallic blue lady beetle (*Curinus coeruleus*) both have with very long branched spines covering their bodies.

Orange-spotted lady beetle (*Brachiacantha ursina*) larvae are unique in that they are found belowground where they live in ant nests and consume scale insects. Larvae of *Scymnus* and *Hyperaspis* are the oddest lady beetle larvae; they are covered with white wax spines.

Photo: Mary M. Gardiner

△ *Many species of lady beetle larvae resemble this multicolored Asian lady beetle (*Harmonia axryidis*): slender and black with short spiders and light markings.*

Photo: Lynette Schimming

△ *Larvae of the twice-stabbed lady beetle (*Chilocorus stigma*) are covered with long, branched spines. The spines likely offer some protection to the larva from potential predators.*

Photo: Abigail M. Parker

△ *The larvae of a few lady beetle species are very unique, with bright white wax covered spines. These* Scymnus *larvae are foraging for aphids.*

A lady beetle was responsible for the first highly successful biological control project (see chapter 2 to learn more about biological control). The vedalia beetle (*Rodolia cardinalis*) was imported in the late 1800s to control the cottony cushion scale, which was a devastating citrus pest. The beetle was incredibly effective and to this day can be found foraging in trees for pests. Since then, several species of lady beetles have been introduced, both intentionally and unintentionally, into the United States. A few of these species have become very common in all or part of the United States, including the seven-spot lady beetle (*Coccinella septempunctata*), multicolored Asian lady beetle (*Harmonia axyridis*), checkerspot lady beetle (*Propylea quatuordecimpunctata*), and variegated lady beetle (*Hippodamia variegata*). These exotic lady beetles do provide pest control in gardens, but their arrival and establishment in the United States also happens to coincide with the decline of several species of native lady beetles within some or all of their former range. These include the nine-spotted lady beetle (*Coccinella novemnotata*), two-spotted lady beetle (*Adalia bipunctata*), and convergent lady beetle (*Hippodamia convergens*). Of the introduced species, the multicolored Asian lady beetle in particular is seen in a negative light by homeowners because it seeks shelter from winter inside homes. It also feeds on grapes late in the growing season, and if harvested with the fruit, the beetle's distasteful flavor can contaminate wine.

80 Good Garden Bugs

Guide to Lady Beetle Identification

Adult lady beetles are identified based on characteristic spots, shapes, or other patterns on their pronotum, a plate covering the thorax of the beetle, and their elytra or wing covers.

Photo: John and Jana Balaban

Fifteen-Spotted Lady Beetle
(*Anatis labiculata*)
Size: 7 to 10 mm
Distinguishing features: Black and white pronotum; elytra darken with age from gray to purple. They have a total of fifteen black spots.
Prey: Aphids in trees

Photo: Thomas Bentley (thomasbentley.com)

Giant Lady Beetle
(*Anatis lecontei*)
Size: 7 to 11 mm
Distinguishing features: Black pronotum with two white stripes; red elytra boarded by a black stripe
Prey: Aphids in trees

Photo: Peter Cristofono

Eye-Spotted Lady Beetle
(*Anatis mali*)
Size: 7 to 10 mm
Distinguishing features: Reddish brown elytra with black spots surrounded by light rings
Prey: Aphids in trees

Photo: Bill Johnson

Orange-Spotted Lady Beetle
(*Brachiacantha ursina*)
Size: 3 to 4 mm
Distinguishing features: Black pronotum with two orange spots; black elytra, each with five orange spots.
Prey: Scale insects and mealybugs

Photo: Charley Eiseman

Twice-Stabbed Lady Beetle
(*Chilocorus stigma*)
Size: 4 to 6 mm
Distinguishing features: Black pronotum; black elytra, each with one red spot
Prey: Aphids and scales in trees

Photo: Libby & Rick Avis

Three-Banded Lady Beetle
(*Coccinella trifasciata*)
Size: 4 to 5 mm
Distinguishing features: Pronotum bordered in white toward head, rest black; yellow to red elytra, each with three long, black bands with light outlining
Prey: Aphids

Photo: Carol Davis

Transverse Lady Beetle
(*Coccinella transversoguttata*)
Size: 5 to 8 mm
Distinguishing features: Black pronotum with two white spots; elytra with solid black band behind pronotum, each with two elongated black spots
Prey: Aphids

Photo: Shutterstock.com

Seven-Spotted Lady Beetle
(*Coccinella septempunctata*)
Size: 7 to 8 mm
Distinguishing features: Black pronotum with two white spots; red elytra, each with three spots and one central spot just behind pronotum
Prey: Aphids

Guide to Lady Beetle Identification (cont'd)

Photo: Libby & Rick Avis

Nine-Spotted Lady Beetle
(*Coccinella novemnotata*)
Size: 5 to 7 mm
Distinguishing features: Two forms, both have a black pronotum with two white spots; one form has red elytra without spots. The other form has red elytra, each with four black spots and one central spot.
Prey: Aphids

Photo: Nolie Schneider

Pink Lady Beetle
(*Coleomegilla maculata*)
Size: 4 to 6 mm
Distinguishing features: Dark pink pronotum with two triangular-shaped black spots; dark pink elytra with multiple black spots.
Prey: Aphids, also commonly feeds on corn pollen

Photo: Matt Edmonds

Metallic Blue Lady Beetle
(*Curinus coeruleus*)
Size: 5 to 6 mm
Distinguishing features: Metallic blue pronotum with two orange spots; metallic blue elytra.
Prey: Scale insects and psyllids

Photo: Brandon Woo

Mealybug Destroyer
(*Cryptolaemus montrouzieri*)
Size: 3 to 4 mm
Distinguishing features: Orange to brown pronotum; elytra black toward pronotum with brown present at tips; dark legs
Prey: Mealybugs

Photo: Beatriz Moisset

Spotless Lady Beetles
(*Cycloneda munda*, *Cycloneda polita*, **and** *Cycloneda sanguinea*)
Size: 4 to 6 mm
Distinguishing features: Three species are found in the United States. They have a black pronotum with a white scalloped edge that creates two black circles on either side. The elytra are bright red without spots.
Prey: Aphids

Photo: Matt Edmonds

Multicolored Asian Lady Beetle
(*Harmonia axyridis*)
Size: 5 to 8 mm
Distinguishing features: There are many variations in color and pattern in this species. The pronotum is always white with a black "W" pattern. The elytra may be yellow, red, or black, with or without spots.
Prey: Aphids

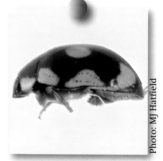

Photo: MJ Hatfield

Undulate Lady Beetle
(*Hyperaspis undulata*)
Size: 2 to 3 mm
Distinguishing features: Black pronotum, bordered with light brown; elytra with a pair of light brown spots and an "undulating" brown stripe along their edges
Prey: Mealybugs and scales

Photo: Libby & Rick Avis

Convergent Lady Beetle
(*Hippodamia convergens*)
Size: 4 to 7 mm
Distinguishing features: Black pronotum with two converging white bars; red elytra with several black spots
Prey: Aphids

Glacial Lady Beetle
(Hippodamia glacialis)
Size: 6 to 8 mm
Distinguishing features: Pronotum with two converging white bars; orange to red elytra with large fused spots at base
Prey: Aphids

Parenthesis Lady Beetle
(Hippodamia parenthesis)
Size: 4 to 6 mm
Distinguishing features: Black pronotum with white border; orange to red elytra, each with a parenthesis-shaped spot.
Prey: Aphids

Thirteen-Spotted Lady Beetle
(Hippodamia tredecimpunctata)
Size: 4 to 7 mm
Distinguishing features: Black pronotum with white border and a black spot on each side; orange to red elytra have a total of thirteen black spots, some may be fused together.
Prey: Aphids

Variegated Lady Beetle
(Hippodamia variegata)
Size: 4 to 5 mm
Distinguishing features: Black pronotum with a white border and two white spots; orange to red elytra with black spots
Prey: Aphids

Ash Gray Lady Beetle
(Olla v-nigrum)
Size: 4 to 7 mm
Distinguishing features: Two different color morphs occur in this species. It may have a gray pronotum and elytra with numerous black spots, or a black pronotum with a white border and black elytra, each with a large red spot.
Prey: Aphids, caterpillar eggs and larvae, and beetle larvae in trees

Striped Lady Beetle
(Paranaemia vittigera)
Size: 5 to 7 mm
Distinguishing features: Dark pink pronotum with two triangular spots; dark pink elytra with black stripes
Prey: Aphids

Checkerspot Lady Beetle
(Propylea quatuordecimpunctata)
Size: 3 to 5 mm
Distinguishing features: Black elytra with yellow square-shaped spots
Prey: Aphids

Vedalia Beetle
(Rodolia cardinalis)
Size: 3 to 5 mm
Distinguishing features: Black pronotum with two red spots; dark red elytra with irregular black spots
Prey: Cottony cushion scale

Fireflies/Lightning Bugs (Lampyridae)

These beetles range from small to large (5 to 20 mm), and males and females can look very different. Males are winged and able to fly long distances, while females may or may not have wings, and those with wings tend to make short flights. Females of the pink glowworm, *Microphotus angustus*, (½ an inch [1.5 cm]) are an example of a wingless species and look very much like a larva. Males of this and many other species look similar with brown, gray, or black with red marks on the pronotum and elytra that are soft and leathery. Their pronotum is large and conceals the head from above, differentiating fireflies from the similar-looking soldier beetles (Cantharidae). Males and females have long, thin antennae. Larvae prey on snails or soft-bodied insects, and adults may also, although not much is known about their foraging behavior.

The common names fireflies and lightning bugs come from the ability of many species to produce light through the oxidation of luciferin, a chemical produced in a bioluminescent organ in the abdomen. The color of the light can vary—some emit a green light, while others give off a yellow or white light. Some larvae are also able to produce light and are called glowworms. The light serves to warn predators that these larvae are not good to eat.

Adults produce light predominately to locate mates. Each species has a characteristic flash pattern. A male will fly and flash his light, and an interested female will flash back. Females in the genus *Photuris* (½ to ¾ of an inch [1.5 to 2 cm]) can even mimic the flash patterns of other species. If a male of another species falls for this trick and approaches a *Photuris* female thinking he has found a possible mate, she will not hesitate to make a meal of him!

Although light production is common among adults in this family, not all are bioluminescent. *Ellychnia* (½ to ¾ of an inch [1.5 to 2 cm]) is the dominant genus in the western United States, and these day-active species do not produce light.

Photo: Debbi Brusco

△ *Females of the pink glowworm,* Microphotus angustus, *look a lot like larvae due to their lack of wings. They are found on the soil surface and emit a bright light at night to attract males.*

Photo: Robyn Waayers

△ *Male fireflies look a bit like some solider beetles, but their pronotum, the plate that covers their thorax, extends to conceal their head from above. Males are winged and able to fly long distances in search of the blinking light of their potential mates.*

Soft-Winged Flower Beetles (Melyridae)

These are small- to medium-size beetles may have narrow, elongated bodies or wedge-shaped bodies. Many are metallic with brightly colored markings. As the name suggests, these beetles have softer elytra than many other beetle families; the elytra and pronotum of these soft-winged beetles are often covered with hairs. Soft-winged flower beetles have long antennae, and in males, the first segments are enlarged and used to grasp female antennae. Some species in the genus *Collops* are important agricultural predators, such as the four-spotted collops, *Collops quadrimaculatus,* and two-lined collops, *Collops vittatus* (both 4 to 6 mm). Adults, which are commonly spotted on flowers, also feed on pollen and nectar.

The four-spotted collops, Collops quadrimaculatus *(1), and the two-lined collops,* Collops vittatus *(2), are important predators of many types of pest insects, including whiteflies (Aleyrodidae), aphids (Aphididae), small caterpillars (Lepidoptera), thrips (Thripidae), lygus bug nymphs (Miridae: Lygus), and insect eggs.*

1

Photo: Ilona L.

2

Photo: Mike Quinn

Rove Beetles (Staphylinidae)

Rove beetles are a large family of elongated beetles that are found in or on the soil or mulch in home gardens. This family of beetles is easily distinguished from others by their short elytra, which does not completely cover their abdomen. They range in size from small to large (1 to 25 mm) and are typically brown or black, although some have bright markings. They have curved mandibles and long antennae that are either threadlike or clubbed. If threatened, rove beetles often turn the tip of their abdomen over their back as if to strike, although they do not have any ability to sting. Larvae look similar to adults, except that they lack wing covers.

Rove beetles may be predatory or scavengers that feed on carrion, fungi, or dung. Some are predatory as adults with larvae that develop as parasitoids. The diet of predators has not been thoroughly studied, but they are known to consume root maggot eggs and larvae (Anthomiidae: *Delia*), mites (Acari), and other soft-bodied insects that live in the soil. Many rove beetles that have been observed in agricultural habitats, including *Paederus* (7 to 9 mm), have highly generalized feeding habits. Some species in the genera *Platydracus* (4 to 15 mm) are reported to feed on armyworms (Noctuidae), and some in the genus *Tachinus* (4 to 6 mm) feed on fly maggots (*Diptera*). Both adults and larvae of the spider mite destroyer (*Oligota oviformis*) feed on two-spotted spider mites and the European red mite (Tetranychidae), which are typically found on strawberry, fruit trees, and ornamental deciduous trees and shrubs. This species is small (2 mm) and black as an adult, while larvae are light pink to orange.

The species *Aleochara bilineata* is natural enemy of root maggots. Introduced to the United States from Europe, it measures 5 to 6 mm and occurs throughout the United States. Its main source of food is cabbage maggots

△ *For some reason, I still remember staring at a box of pinned specimens in my undergraduate Entomology laboratory practical trying to decide which insects were beetles. Most were easy, but I stared for a long time at a pinned Staphylinidae and incorrectly decided not to list it. They really don't look much like other beetles. They do have elytra (or hardened wing covers), but they are very short. In this* Paederus *species, the elytra are a metallic blue, and you can see that they only cover about one-quarter of the abdomen. They have a pair of membranous wings folded under their elytra and can fly.*

Photo: Brandon Woo

◁ *Here is a rove beetle in the genus* Platydracus, *illustrating its defensive posture. Although it looks as if it is ready to strike like a scorpion, it's all show. Rove beetles do not have the ability to sting.*

(Anthomyiidae: *Delia radicum*). Females lay their eggs near the base of maggot-infested plants. Larvae hatch and seek cabbage maggot pupae in the soil and will complete their larval development by feeding on one pupa. These larvae will pupate within the remaining shell of the maggot pupa once it has consumed it, eventually emerging as an adult. Adults are predatory of eggs and young cabbage maggots. Rove beetles in the genus *Stenus* (4 to 6 mm) are a very unique adaptation — if they are knocked into a pond or stream, they can release a chemical from their abdomen that lowers the surface tension of the water and causes the beetle to be propelled forward!

Photo: Brandon Woo

△ *Beetles in the genus* Stenus *have a unique defensive strategy: They are able to release an alkaloid from their abdomen to propel themselves forward.*

Beetles in Your Garden Pond

Garden ponds and other water features can be colonized by predatory beetles that are both fun to observe and provide biological control of aquatic pests. These beetles are found in water as both larvae and adults and are predators though both stages. Adults have strong wings and are able to move among bodies of water, which is how they arrive within newly established ponds. They are also attracted to light and are sometimes found on driveways in the morning, under streetlights after dark, or on porches. Even out of their element, their body structure gives them away as aquatic. With legs that look like a paddle of thick bristles, these beetles are modified for swimming.

Predaceous Diving Beetles (Dytiscidae)

Dytiscidae vary from small to large (2 to 35 mm), oval and flattened in shape, and can be brown or black, often with yellowish markings. These powerful swimmers have legs that are modified for swimming and lined with hairs. They use their oarlike hind legs in synchrony to propel themselves through the water of ponds and lakes. Some of the larger species are in the genera *Dytiscus* and *Cybister*, such as the vertical diving beetle (*Dytiscus verticalis*, 1¼ inches [3.5 cm]) and giant diving beetle (*Cybister fimbriolatus*, 1" to 1¼" [2.5 to 3 cm]). Adults will keep a bubble of air under their elytra to allow them to dive for prey while hunting.

Diving beetles sit with their heads below the surface in search of prey and consume many aquatic organisms. After mating, females lay eggs one at a time on aquatic plants. The larvae are called water tigers, and they prey on many arthropods, tadpoles, and even small fish. They have a large head with curved mandibles that serve two purposes: They allow the larvae to inject their prey with saliva, which aids digestion, and they enable the larva to consume a liquid diet. Some water tigers breathe above the water, while others have gills and can remain submerged. They leave the water to pupate in muddy soil at the water's edge.

Photo: Charley Eiseman

△ *The vertical diving beetle,* Dytiscus verticalis, *is black with a yellow band along the edge of its pronotum, the plate that covers the thorax, and its elytra or wing covers. They feed on mosquito larvae, other aquatic insects, and even tadpoles or small fish.*

Photo: Micki Killoran

△ *The larvae of predacious diving beetles, called water tigers, may actively swim around in search of prey and will use their large mandibles to grasp any passing arthropod. These larvae breathe through spiracles located at the tip of their abdomen; they will hang just below the water's surface with it exposed to replenish their air supply.*

Whirligig Beetles (Gyrinidae)

Whirligig beetles (3 to 15 mm) have oval-shaped, flattened bodies that are either black or metallic. They have long raptorial front legs for grasping prey and modified hind legs for swimming. They are often found in groups on the surface of the water. Whirligig beetles are named for their tendency to swim in circles. They produce a defensive secretion from their prothorax, and it is the foul odor of this secretion that gives whirligig beetles the nicknames apple smellers and mallow bugs. The adults have eyes that are divided, allowing them to see above and below the water's surface. When hunting, they hold their club-shaped antennae on the water's surface to detect struggling insect prey. Females deposit their eggs on aquatic plants. These eggs hatch into long and slender larvae with overlapping plates along the dorsal side of their pronotum and abdomen. The larvae are predators and can be found on the bottom of ponds.

Photo: Gayle and Jeanell Strickland

△ *Whirligig beetle adults have divided eyes, giving them the amazing ability to swim at the water's surface while keeping watch below and above water.*

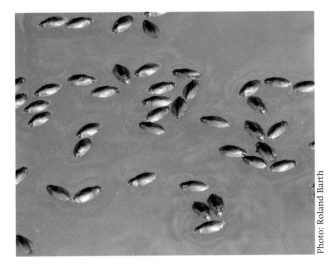

Photo: Roland Barth

△ *Whirligig beetles (Gyrinidae) are often seen swimming on the surface of the water in groups, but if they feel threatened, they will dive underwater.*

8

Predator and Parasitoid Flies (Diptera)

Although some species can be an incredible annoyance, there are many very beneficial flies found in the garden. Some flies act as predators, parasitoids, or even pollinators! Even more amazing, some of the species discussed in this chapter provide two of these important services during their lifecycle. In this chapter, you will be introduced to the species likely to be feeding on insect pests in your backyard. I encourage you to take a closer at this diverse group of insects; their unique beauty and beneficial activity may just change your opinion of flies and their place in the garden.

The order name Diptera means two wings, indicating that each fly in this order has one pair of wings. If they will remain still long enough to allow a careful look, this key characteristic makes it relatively easy to distinguish flies from insects in other orders. The second pair of wings are reduced to small organs called halters that help with balance in flight. Most adult flies also have large eyes and mouthparts that look either like a piercing beak or like a sponge used for lapping up liquid meals. Fly larvae, also called maggots, are legless, but they are still able to move effectively using turgor pressure and muscle action.

These movements can also be aided by small creeping welts or prolegs, which are types of small appendages that help in their movement. Some fly larvae do not have eyes, while others have small eyes called stemmata, which are light-sensitive cells. Flies have complete metamorphosis, meaning they develop through egg, larval, pupal, and adult stages.

Photo: Gayle and Jeanell Strickland

△ Robber flies (Asilidae) have a rigid mouthpart that points down from their head. Some species, such as those in the genus Megaphorus, are covered with dense bristles.

Predatory Flies

As adults and/or larvae, predatory flies may feed on garden pests. All adults and some larvae feed above ground and attack a wide range of prey, including aphids. The larvae of many predatory flies live below ground in moist soils. Little is known about the diet of these species, but it likely includes a diversity of soil-dwelling invertebrates. This chapter examines seven families of predatory flies that may be found hunting in home landscapes.

Robber Flies (Asilidae)

Robber flies range from small to large (3 to 50 mm), with a characteristic "beard" of hairs surrounding their mouthparts, which are rigid and point down and slightly forward from their head. These flies can inflict a painful bite, so be careful when handling them! Their bodies may be covered with dense bristles, such as *Megaphorus*, or nearly hairless, such as the hanging-thieves robber flies (*Diogmites*) and *Atomosia* species.

Robber flies may be long and slender like the *Neoitamus* or *Leptogaster*, or robust like the giant robber flies (*Promachus*), which measure ¾ to 1¼ inches (2 to 3 cm) and have a striped pattern on their abdomen. In all cases, robber flies will have a thorax that is wider than their abdomen. They vary in color from gray to brown or black, and some will have bright markings. For example, the bee killers (*Mallophora*) and beelike robber flies (*Laphria*) are bee mimics with bright patches of orange, yellow, or white hairs on their bodies. These species do feed on bees and wasps (Hymenoptera), but also on many other types of insects. Those in the genus *Ceraturgus* are wasp mimics with few hairs and characteristic black bodies with yellow markings on their abdomen.

△ *Robber flies do not feed solely on insect pests. The bee killer robber flies (*Mallophora*) are bee mimics that feed on beneficial bees, wasps, and other insects.*

▷ *Hanging-thieves robber flies (*Diogmites*) are nearly hairless with long, slender legs. Often they will use their forelegs to hang from a perch while consuming their prey, a behavior that led to their common name.*

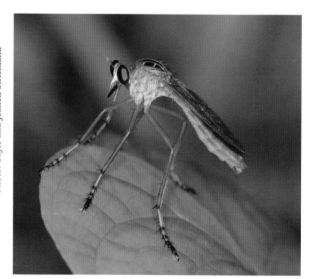

Photo: Gayle and Jeanell Strickland

Photo: Matt Edmonds

All adult robber flies are voracious predators that feed on various types of insects, including other beneficial insects and many that are considered pests. Adults catch their prey on the wing, grasping it with their legs and then finding a good location to land and consume it. Robber flies inject their prey with saliva, which includes enzymes that rapidly paralyze their victim and begin to break down its internal contents. The fly then consumes its prey as a liquid meal.

Robber flies are most vulnerable to predation when mating. The male and female often face away from each other while mating and although not impossible, flight is limited at this time.

After mating, females lay eggs in soil or on plants or even drop them during flight. Larvae are light in color with dark mouthparts and a segmented body. Some species have rounded turbicals protruding from their body segments. The predatory behavior of the larvae is not well understood, but most forage for prey in moist soil, under bark, or within decaying wood. Robber flies will also pupate within these same protected locations.

△ *Giant robber flies* (Promachus), *which measure ¾ to 1¼ inches (2 to 3 cm), are among the largest species in the family.*

Photo: Patrick Coin

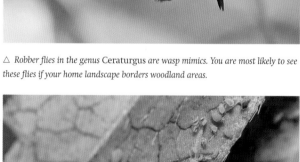

△ *Robber flies in the genus* Ceraturgus *are wasp mimics. You are most likely to see these flies if your home landscape borders woodland areas.*

Photo: Steve Collins

▷ *Larvae of the aphid predatory midge* Aphidoletes aphidimyza *are eyeless maggots. When they come in contact with the leg of their aphid prey, they bite it to paralyze the aphid. They then ingest the liquid contents of the prey, leaving a shriveled carcass, several of which are visible surrounding the larva in this photo.*

Photo: Dan Leeder

Predator and Parasitoid Flies (Diptera) 93

Predatory Midges (Cecidomyiidae)

The family Cecidomyiidae is composed mainly of flies that feed within plant parts, causing the formation of galls – deformations of plant tissue that these insects can live within and feed on as larvae. The Cecidomyiidae also include some predatory species that do not form galls. One important species is the aphid predatory midge, *Aphidoletes aphidimyza*. This fly is a minute (2 mm), delicate midge with long legs and slender antennae. The adults are rarely seen because they are active at night, feeding on honeydew, a sweet sap that aphids (Aphididae) secrete

Photo: Troy Bartlett

△ *Adults of the aphid predatory midge* Aphidoletes aphidimyza *feed on aphid honeydew, a sugar-rich secretion the aphids produce as they feed on plants. Male and female midges mate while hanging from spider webs.*

as they eat. This predator has a very interesting mating behavior. The female finds a spider web, hangs from it, and emits a pheromone from her front legs, which attracts males to the area. Mating occurs with the pair hanging from the web. Perhaps the spider offers the mating pair some protection from other predators, provided the spider doesn't decide to eat the midges itself!

Mated females deposit tiny orange eggs within aphid colonies. They hatch into orange to red larvae. Often, many larvae can be found on a single aphid-infested leaf. The larvae attack by piercing the leg of an aphid and paralyzing it. They then feed on the liquid contents of their prey, leaving a shriveled carcass behind. When ready to pupate, the larvae drop from plants and pupate in a cocoon they spin in the soil. The overwintering generation will spend the winter within this cocoon before emerging as adults in the spring.

The predatory midge *Feltiella acarisuga* is another important predator that feeds on spider mites (Tetranychidae) throughout the United States. This midge looks very similar to the aphid predatory midge, but you will find the larvae of this species on spider mite–infested leaves instead of the aphid hunting grounds preferred by the aphid predatory midge. Adults are minute (2 mm) with long legs and a pinkish brown body. They do not feed on prey during the adult stage. After mating, females lay oval-shaped, translucent white eggs on spider mite–infested plants. The eggs hatch into orange larvae, which feed exclusively on spider mites, sometimes consuming more than 300 spider mites and mite eggs before completing their development cycle.

Long-Legged Flies (Dolichopodidae)

The Dolichopodidae are very commonly found resting on plant leaves in the garden. The adults are small flies (1 to 9 mm), with long, delicate legs. Many, such as *Condylostylus*, have metallic green, blue, or copper bodies. Others, such as *Xanthochlorus* and some *Gymnopternus* species, are yellow, brown, or black. Some long-legged flies have clear wings, and others have black patterning on their wings.

Unlike dance flies or robber flies, which have an obvious piercing beak, it is difficult to see the mouthparts of a long-legged fly without the aid of a microscope. Nearly all species of long-legged flies are predatory as adults. They feed by grasping at their prey with sawlike mouthparts, which are covered with soft structures that allow the flies to lap up their meal. Long-legged flies feed on a diversity of small, soft-bodied arthropods such as springtails, aphids, and other flies, along with many insect eggs. In the genus *Medetera*, larvae are known to attack bark beetles specifically.

In some species, such as those in the genus *Dolichopus*, males perform mating dances to attract females using large flattened scales on their front legs that they wave like flags to draw in potential mates.

Mated females lay eggs that hatch into translucent, white smooth-bodied larvae. Very little is known about the larvae of long-legged flies, however, nearly all species are predatory, and they can be found in moist soil or under bark.

▷ *Male long-legged flies in the genus* Dolichopus *perform mating dances by waving the scales on their front legs.*

Photo: Roland Barth

△ *Many long-legged flies (Dolichopodidae) are metallic colored, including this one in the genus* Condylostylus.

Photo: Dennis Gauthier

△ *Not all Dolichopodidae are shiny and metallic, but this individual is showing off the long legs common to members of this predatory fly family.*

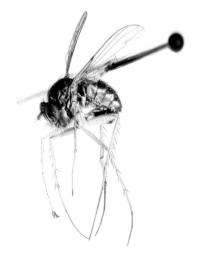

Photo: Andrea Kautz

Dance Flies (Empididae)

Dance flies are minute to medium-size (2 to 15 mm) and look somewhat similar to robber flies (Asilidae). However, dance flies do not have a "beard" of bristles; instead, they have a very round head and mouthparts that point downward or backward from their head. Most dance flies are fairly drab in color, with grey, tan, or black bodies. In some species, such as the balloon flies (*Hilara*), males have an enlarged segment on their front legs; in others, such as the long-tailed dance fly (*Rhamphomyia longicauda*), females have fringed legs lined with long bristles.

Dance flies are predatory as adults, but they can often be found on flowers, feeding on pollen and nectar as well. The larvae also hunt prey and may be found in the water, in moist soil, or under the bark of trees. Dance flies have a very unique mating behavior. Males kill and wrap an insect prey in a silken material they produce in their front legs. This is provided as a "gift" to a female prior to mating. In many species, males form swarms, with each individual holding its wrapped prey; females will fly into the swarm to select a mate. In some species, however, the females form mating swarms, and males fly into it with their insect prey gifts. Sometimes, males do try to deceive the females by creating an empty balloon of silk as their "prey" offering.

△ *A female long-tailed dance fly,* Rhamphomyia longicauda, *shows off her fringed legs.*

Photo: D. Shetlar

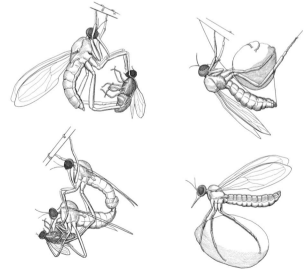

△ *Male dance flies go through a lot of effort to attract the perfect mate. First they catch and kill an insect and wrap up this "gift" in a silken material. Male dance flies then take flight, joining other males to display their wrapped insect pests to prospective mates. Females consume this nuptial gift during mating.*

Snipe Flies (Rhagionidae)

Snipe flies are robust, small- to medium-size flies (4 to 15 mm) with pointed abdomens that look somewhat similar to horse flies or deer flies. They are typically dull yellow or brown, but some, such as the gold-backed snipe fly (*Chrysopilus thoracicus*) and ornate snipe fly (*Chrysopilus ornatus*), have gold metallic hairs on their thorax and abdomen. They may have clear wings, patterned wings like the common snipe fly (*Rhagio mystaceus*), or they may have markings on their wings like the small flecked-wing snipe fly (*Rhagio lineola*). As adults, some snipe flies are predatory, but many feed on pollen and nectar. Some species will rest on tree trunks in a characteristic stance with their head pointed down, earning them the name downlooker flies. The larvae of snipe flies are predators and can be found in decaying wood, garden soil, and compost.

Photo: D. Shetlar

△ *The gold-backed snipe fly* (Chrysopilus thoracicus) *is a lovely species, with gold hairs on its thorax and abdomen.*

Photo: Nolie Schneider

▷ *Snipe flies can have clear wings, patterned wings, or wings with markings such as the flecked wing snipe fly* (Rhagio lineola). *All species have fairly large eyes and a pointed abdomen.*

△ *Snail-killing fly adults are yellow, tan, or brown flies with spotting wings, such as this individual in the genus* **Tetanocera**.

△ *Here is a snail-killing fly larva in the genus* Sepedon. *These flies attack aquatic snails and can be found just below the surface of the water.*

Snail-Killing Flies (Sciomyzidae)

Snail-killing flies are small- to medium-size (2 to 12 mm) yellow, tan, or brown flies that typically have spotted wings and antennae that point forward. Given their common name, it's not surprising that the larvae are predators or parasitoids of mollusks. The adult flies are found in habitats where their prey are common, including the edges of ponds and streams as well as garden habitats frequented by snails and slugs. Sciomyzidae larvae, such as those in the genus *Sepedon*, attack aquatic snails and can be found just below the surface of a pond or lake. They are able to float, suspended by water-repellent hairs, in the surface film of the water, obtaining oxygen though a breathing tube that maintains contact with the open air. When a Sciomyzidae larva encounters a snail, it attacks the snail's exposed foot, and as the snail retracts, the predator is pulled inside the shell, where it kills and consumes its prey. Other genera, including *Tetanocera*, feed on both aquatic and terrestrial snails and slugs. Larvae that feed on slugs develop as parasitoids, entering the slug though the mouth, eye stalks, or mantel, and develop within it, eventually causing death.

Hover Flies or Flower Flies (Syrphidae)

Syrphidae are one of the most common and attractive beneficial flies you will find in the garden. Their name comes from their ability to hover in midair and make quick motions while darting among flowers hunting for prey. Hover flies are small to large (4 to 24 mm) and often resemble bees or wasps. As adults, these flies are pollinators, and they feed on pollen and nectar. Some genera, such as *Eristalis*, are hairy and beelike, with dark bodies and light-colored strips on their abdomen. Other species, such as those in the genus *Toxomerus* and the four-spotted aphid fly (*Dioprosopa clavata*), lack hair and more closely resemble wasps.

△ *Many hover flies, such as those in the genus* Toxomerus, *are wasp mimics and lack hair.*

Predatory hover fly females will lay oval-shaped eggs within areas inhabited by prey, including aphids (Aphididae) and other soft-bodied garden pests such as whiteflies (Aleyrodidae) and scales (Coccoidea). Eggs hatch into eyeless maggots that look somewhat sluglike. The larvae are often green or brown with a relatively smooth-looking body or one with rigid projections along it. Adult females of these species lay single eggs or clusters of eggs (depending on the species) on garden plants, near or within pest infestations.

▷ *Some hover flies, such as those in the genus* Eristalis, *are hairy and beelike.*

△ *Hover fly larvae are often green or brown, and they may appear relatively smooth or have ridges on their bodies.*

Parasitoid Flies

Parasitoid flies have a specialized life cycle that differs from a parasite in that the host is killed by the fly during its development. Female parasitoids search for a suitable host to support the development of her offspring, which will feed as larvae on this host, eventually killing it.

Some parasitoids attack a specific species of arthropod, while others can develop within several possible hosts. And some female fly parasitoids may lay eggs inside the host insect, where larvae will hatch and begin feeding on the host's internal organs and tissues. Other female parasitoids lay their eggs on the "skin" or exoskeleton of the host insect where the larvae will hatch and burrow into the host before feeding. Females may also lay eggs on plant material, where they will be eaten by the host. The eggs that are not destroyed during feeding will hatch within the gut of the host. In other species, females will lay eggs throughout the environment, and the larvae are left to find their own suitable host.

Parasitoid Flies (Tachinidae)

The family Tachinidae is one of the largest in the Diptera, with more than 10,000 species. These flies range from 1 to 17 mm in length and come in a wide variety of colors — many are dark and resemble house flies, but some have brightly colored markings. Their thorax and abdomen are covered in thick bristles. Adults feed on pollen and nectar and are commonly seen foraging on garden flowers. Females may lay eggs in or on hosts, or in their environment, depending on the species of Tachinidae. The larvae will develop internally within the host, eventually killing it. Most parasitoid flies will emerge from the host to

◁ *The feather-legged fly* Tricopoda pennipes *is a parasitoid of the squash bug. This fly has a bright orange abdomen and featherlike scales lining its hind legs.*

Photo: Ben W. Phillips

◁ Hystricia abrupta *has a robust, bright-orange abdomen with prominent dark bristles.*

Photo: Tom Bentley

pupate, although some will pupate within it. The adult fly will emerge from the pupa and search for a mate.

Tachinidae are one of several insects that exhibit an interesting courtship behavior called hilltopping, where males and virgin females fly to high points within a landscape to mate. A large proportion of the flies on the hilltop will be males that perch on the branches of shrubs and trees. They often have a favorite perch or territory, and they will fly out from their perches when other insects come near. Males will sometimes show aggression, chasing other males out of their territory. Virgin females will select a male, and after mating, they will leave the hilltop, never to return.

Many Tachinidae are important parasitoids of garden pests. One very colorful example is the feather-legged fly, *Trichopoda pennipes*, which attacks squash bugs (Coreidae: *Anasa*). This fly has a bright orange abdomen and feather-like scale projections on its hind legs. Other important garden-dwelling parasitoid flies include the less charismatic *Celatoria diabroticae* and *Celatoria compressa*. These flies attack cucumber beetles (Chrysomelidae: *Acalymma vittatum* and *Diabrotica undecimpunctata*), which can be devastating garden pests. Other Tachinidae, such as *Voria ruralis* and *Hystricia abrupta*, attack common caterpillar pests.

Photo: Ben W. Phillips

◁ *Two Tachinidae in the genus* Celatoria *attack cucumber beetles in the garden. Here a pupa of a* Celatoria *is seen bursting out of a dead striped cucumber beetle.*

The Life Cycle of the Feather-Legged Fly
(Trichopoda pennipes)

△ *A mated female feather-legged fly locates a squash bug in the garden.*

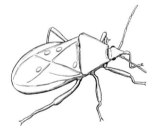

△ *The female fly will deposit one or more eggs on the pest.*

△ *The adult feather-legged fly will emerge from the pupal case.*

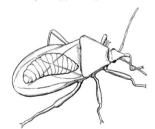

△ *Larvae will hatch from each egg and burrow directly into the body of the squash bug. Only one larva will survive.*

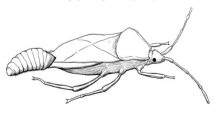

△ *When the larva is ready to pupate, it emerges from the dead squash bug and pupates in the soil.*

9

Wasps and Ants (Hymenoptera)

A stroll down the pest-control aisle of any home improvement store makes it clear how many of us feel about wasps and ants. Some in the order Hymenoptera certainly can be household pests or even a safety concern, but on the whole, the good these insects do in terms of pest suppression far outweighs the bad. And many wasps and ants are strikingly beautiful and exhibit fascinating behaviors. I hope learning more about them will make these creatures a more intriguing and less frightening part of your home landscape.

The Hymenoptera is considered the third largest order of insects, but it may actually be the largest order of insects (a title currently held by beetles). Many species in this family are incredibly tiny, and it's likely that there are many new species yet to be discovered. Wasps and ants develop by complete metamorphosis, meaning that they have an egg, larval, pupal, and adult stage. In this chapter, I will discuss the beneficial species of the order Hymenoptera, which fall within three groups: parasitoid wasps, stinging wasps, and ants.

Parasitoid Wasps

Parasitoid wasps have a very interesting life cycle. A parasitoid is similar to a parasite in that it relies on a host animal to survive. Unlike a parasite, however, a parasitoid eventually kills the host as part of its development. Parasitoid females also spend a good portion of their time looking for insects to serve as food for their offspring. They may seek out one particular species of arthropod or be less selective and accept a number of different hosts, usually in the same family or at least order. Parasitoid wasps usually attack a particular life stage of their host, but among different wasp species, all life stages — egg larvae or nymph, pupa, and adult — are attacked. Females use visual cues as well as odors and/or sounds to locate an arthropod host, which may be hidden in a rolled leaf, inside a plant stem, or even under bark. Once a suitable host is found, the female will insert her ovipositor, or egg-laying organ, into the host and deposit one or more eggs inside. The ovipositor can be very short and difficult to see or very long as in some Ichneumonidae species. Usually each egg hatches into an individual larva that will begin feeding internally on the host, but in some cases, in a process called polyembryony, females can insert one egg that divides, resulting in multiple larvae. In some species, the host ceases development when parasitoid larvae are present, and in other cases, the host continues to develop and dies later, usually when the wasp enters the pupal stage. The wasp may pupate inside the host, attached to the exoskeleton, or leave the host entirely to pupate. When they emerge as adult males and females, parasitoids will feed on pollen and nectar and are often found in flowers. Sometimes adult females will feed on the species of hosts they seek for their offspring as well.

It's important to note that only female wasps will have an ovipositor, which cannot be used to sting people, so although some parasitoids may look frightening, they pose no threat to humans.

Braconid Wasps (Braconidae)

Braconidae is a large and diverse family of minute- to medium-size wasps (1 to 18 mm). Some braconid wasps look similar to those in the family Ichneumonidae. If you are able to inspect them, Braconidae wings will not have a "horse head" shape, which is a key identification characteristic for Ichneumonidae. Also, instead of pupating in their host (as many Ichneumonidae do), Braconidae species often construct a silken pupae on their host or pupate away from their host entirely.

Many braconid wasps attack caterpillars with a range of feeding strategies. For example, the genera *Mirax* (1 to 3 mm) parasitizes tiny leafminers. Others, including *Cotesia* (5 to 8 mm), attack larger caterpillars such as armyworms (Noctuidae), cabbage looper (Noctuidae: *Trichoplusia ni*), corn earworm (Noctuidae: *Helicoverpa zea*), diamondback moth (Plutellidae: *Plutella xylostella*), gypsy moth (Erebidae: *Lymantria dispar*), and hornworms (Sphingidae). Several of these small wasps can develop in one host. When ready to pupate, they will emerge and construct a silken cocoon, which some species will attach to their dead or dying host. Not all braconid wasps attack caterpillars. *Peristenus digoneutis* (2 to 5 mm) parasitizes the tarnished plant bugs (Miridae: *Lygus* spp.). Larvae of this small, dark-colored wasp (3 mm) develop in nymphs of plant bugs, and when ready to pupate, they leave the host, which kills it, and construct a cocoon in the soil. Braconid wasps such *Microctonus vittatae* (2 to 3 mm) are important in controlling beetle pests. This wasp utilizes flea beetle (Chrysomelidae) adults as hosts, and the larva sterilizes the female as it develops, preventing the pest

△ *Braconid wasps range from minute to medium-size. The tiny* Mirax, *commonly found in trees and shrubs, attacks insects that create mines in leaves.*

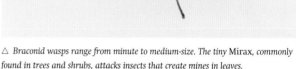

△ Cotesia *attack many common caterpillar pests in the garden. You may not see the small, dark adult wasps, but you are more likely to find the silken white cocoons they spin on the outside of the parasitized caterpillar.*

from laying eggs. This parasitoid can produce female offspring without mating, thus males are very rare!

Many genera of braconid wasps, including species of *Aphidius*, *Praon*, *Lysiphlebus*, and *Diaeretiella* (all less than 5 mm), are important aphid (Aphididae) parasitoids. Adult females search for aphid infestations and, when located, are able to bend their abdomen underneath their thorax and move forward, stinging the aphids and depositing one egg inside each individual. The eggs will hatch, and the larva will consume the contents of the aphid, killing it. When the wasp is ready to pupate, the aphid turns hard and brown and serves as a pupal case for the wasp. At this stage, it is referred to as an aphid mummy. When ready to emerge, the wasp chews an exit hole in the dead aphid. You might notice one or more of these dark aphids within a patch of seemingly healthy ones. It is highly likely that others have been stung, and if left unsprayed for a few days, the parasitoid can develop, emerge, and attack additional aphid pests.

Photo: Stan Gilliam

△ *The yellow and black in the genus* Conura *looks a bit like a yellow jacket* (Vespula), *but its large hind legs give it away.*

Chalcid Wasps (Chalcididae)

Chalcididae are an interesting group of parasitoids that measure 2 to 18 mm and are usually bicolored, although rare metallic Chalcididae have been found. What sets them apart are their enlarged hind legs. Some species will hold the host's jaws open with their strong back legs and deposit their eggs inside the oral cavity of the insect! A few species are also known to fight over desirable hosts, sometimes kicking at each other. Chalcid wasps attack caterpillars (Lepidoptera), beetles (Chrysomelidae), and flies (Diptera), among many others.

Ichneumon Wasps (Ichneumonidae)

The Ichneumon wasps are a large and diverse family, with thousands of species present within the United States. They range in body length from small to large (3 to 50 mm), not including the ovipositor in females. They are typically slender and range substantially in color and pattern. Ichneumon wasps have a particular wing characteristic that is diagnostic to the family; a "horse head" shape is present in their forewing, but it can be

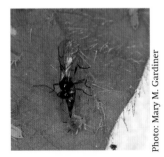

Photo: Mary M. Gardiner

◁ *Some female parasitoids are able to bend their abdomen at their narrow "waist" and use it like a sword to inject eggs into their prey. This wasp is depositing one egg inside an aphid pest.*

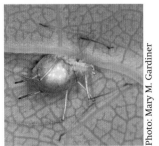

Photo: Mary M. Gardiner

◁ *This aphid mummy has been killed by a parasitoid wasp. The hardened exoskeleton of the dead pest serves as a pupal case for the developing natural enemy.*

very difficult to see on a live wasp. If you see a large, slender wasp with a long ovipositor and long antennae in your garden, there is a good chance it is a species of Ichneumonidae; however, they do have relatives that look quite similar. Ichneumonidae attack a diversity of hosts, including larvae and pupae of beetles (Coleoptera), caterpillars of moths and butterflies (Lepidoptera), and sawflies (Hymenoptera and Symphyta).

Although many Ichneumonidae have a long ovipositor, the short-tailed ichneumon wasps (*Ophion*, ½ to ¾ of an inch [1 to 2 cm]) are among some of the most common species found within the family. These wasps have a slender abdomen that is compressed from front to back. They are typically yellow to reddish brown and have a very short ovipositor that is barely visible. Most short-tailed ichneumon wasps attack caterpillar species. Wasps

in other genera such as the species *Diadegma insulare* (6 mm) are also parasitoids of caterpillars. This parasitoid attacks the diamondback moth in its larval stage and pupates inside the cocoon made by the mature pest larva. If you find diamondback moth pupae on garden plants and see a dark body inside, leave it; a parasitoid and not a pest will emerge. Other ichneumon wasps that attack vegetable garden pests include *Eriborus terebrans*, a black wasp (6 to 10 mm) with reddish-brown legs that attacks the European corn borer, and *Hyposoter exiguae*, a parasitoid of many common caterpillar pests, including armyworms (Noctuidae), cabbage looper (Noctuidae: *Trichoplusia ni*), hornworms (Sphingidae), and tomato fruitworm (Noctuidae: *Helicoverpa zea*). The adult *Hyposoter exiguae* has a dark thorax and head with an orange abdomen and measures 6 to 12 mm in length. The larva

△ *Most short-tailed ichneumon wasps in the genus* Ophion *attack caterpillars. One larva develops per caterpillar. The host is dead when the* Ophion *is ready to pupate, which occurs within the host's body. You can see the horse-head shape common to Ichneumonidae in this individual's wing, just above its front leg.*

△ *This speices of* Diadegma *illustrates the long slender body common to Ichneumonidae. The speices* Diadegma insulare *is a parasitoid of the diamondback moth (Plutellidae:* Plutella xylostella*), a major garden pest of cabbage and other cruciferous crops.*

△ *When the larvae of* Hyposoter exiguae *are ready to pupate, they leave the host and construct an oval-shaped black and white pupa that you may find attached to plant leaves or stems.*

△ *Wasps in the genus* Hyposoter *attack several caterpillar pests in the garden.*

1

2

△ *Several Ichneumonidae attack tree pests including these wasps in the genera* Campoplex *(1) and* Glypta *(2).*

of this parasitoid consumes its caterpillar host and then pupates inside the larval skin or constructs a characteristic pupa outside of the host that is attached to vegetation. *Charops annulipes* (7 mm) constructs a similar pupa and attacks the green clover worm (Erebidae: *Hypena scabra*).

Ichneumon wasps are also important in the control of many tree pests, including larvae that feed on leaves as well as those that feed internally within the wood. For example, *Lathrolestes nigricollis* (3 to 5 mm) is an important parasitoid of the birch leafminer (Tenthredinidae: *Fenusa pumila*), a sawfly that feeds within leaves of birch trees in the northern United States, causing the canopy to brown and affecting tree growth as well as aesthetic appearance. Evergreen trees can also benefit from the activity of ichneumon wasps; *Campoplex frustranae* (3 to 5 mm) and *Glypta fumiferanae* (5 to 8 mm) attack pine tip moths (*Rhyacionia*) and spruce budworm (Tortricidae: *Choristoneura fumiferana*), respectively.

The Hidden Wasps

△ *A few years ago when visiting my parents for the Fourth of July, I heard a commotion outside and found my family backing slowly away from the grill next to a large tree. The tree was under attack by horntail sawfly larvae and a* Megarhyssa *female was busy parasitizing them! After assurance that this large wasp could not sting anyone with its very long ovipositor, the grilling continued.*

The largest parasitoids that may be seen in a home landscape are the giant ichneumons (*Megarhyssa*), which range from ¾ to 2¾ inches (2 to 7 cm) in length, with females measuring up 4½ inches (11.5 cm) with their ovipositor. These large wasps use their long ovipositor to parasitize horntail sawfly larvae (Siricidae) that feed inside dead or dying trees and stumps. Females are able to detect the larvae inside the tree and insert their ovipositor to deposit an egg into the host larva. The ichneumon larva will consume the host and pupate in its burrow. Adults then emerge by chewing an exit hole out of the tree. Emerged males will detect females still in their exit holes and will wait at this location to mate, which typically takes place shortly after her emergence. In some instances, however, males mate with females when they are still in the process of emerging from the exit holes.

Although most Ichneumonidae are endoparasitoids, meaning their larvae develop inside of a host, there are a few ectoparasitoids where the larva of the wasp feeds externally on its host arthropod. One example of this is the species *Sinarachna pallipes*, (5 to 7 mm) which attacks spiders.

Beyond the braconid, chalcid, and ichneumon wasps, which in many cases are large enough to see, a vast diversity of much smaller parasitoid wasps are busy attacking garden pests. Some of the tiniest parasitoids, including the fairy flies in the family Mymaridae and the Trichogrammatidae, attack insect eggs. Mymaridae, which measure 1 to 2 mm in length, get their name from their beautifully fringed wings. Trichogrammatidae are minute, less than 1 mm in length, and are usually yellow or tan. Both families attack eggs from several insect orders. Other hidden wasps include Eulophidae, which are minute to small (1 to 6 mm) in length with slender bodies that vary in color but are often metallic. These wasps often attack larvae or pupae that live in protected areas such as leaf or stem mines. Aphelinidae and Platygastridae range in size from 1 to 3 mm and commonly attack scales (Coccidae), whiteflies (Aleyrodidae), aphids (Aphididae), or psyllids (Psyllidae).

Other families have species that range from minute to just very small. You might find an Encyrtidae (0.5 to 5 mm) attacking scales (Coccidae), aphids (Aphididae), caterpillars, or psyllids (Psyllidae). Pteromalidae are 1 to 5 mm and typically metallic green or blue. They may develop as individual larva or as several larvae within one host insect. They attack a large diversity of insects including scales (Coccidae), aphids (Aphididae), flies (Diptera), and beetles (Coleoptera). Eurytomidae range in size from 1 to 7 mm and are typically brown or black, sometimes with light markings. Many attack insects in protected feeding locations such as wood-boring beetles and insects that form plant galls.

The Hidden Wasps

Here's a closer look at the tiny parasitoids that forage virtually unnoticed within the garden. Avoiding the use of pesticides and providing pollen and nectar plants for these and other wasps to feed on will increase their abundance and contributions to pest management.

Photo: Scott Justis

Encyrtidae
Size: 1 to 5 mm

Photo: Scott Justis

Eulophidae
Size: 1 to 6 mm

Photo: Gary McDonald

Eurytomidae
Size: 1 to 7 mm

Photo: MJ Hatfield

Mymaridae
Size: 1 to 2 mm

Photo: Charley Eiseman

Platygastridae
Size: 1 to 3 mm

Photo: Stephen Luk

Pteromalidae
Size: 1 to 5 mm

Photo: Scott Justis

Trichogrammatidae
Size: less than 1 mm

The Stinging Wasps

The stinging wasps are predatory and collect insect prey to feed their young. All of the female wasps within this group have an ovipositor that is modified to sting, which they use to paralyze prey and/or to defend themselves or their nest. Male wasps do not sting. Some families within this group paralyze prey in protected sites and lay an egg onto the prey. The larva then develops by feeding on this immobile insect. Others transport paralyzed prey to a nest they construct. Stinging wasps may live a solitary or social life.

Bethylid Wasps (Bethylidae)

Bethylidae are small (3 to 6 mm), slender wasps that are typically tan, brown, or black. Females find caterpillars or beetle larvae in protected sites, such as rolled leaves. She will sting the insect to paralyze it and then carefully chew

it without rupturing the prey's skin or cuticle so that her offspring can easily consume it. The female will then lay several eggs on this caterpillar, which will hatch and develop by consuming their tenderized insect meal. Eventually, the larvae will pupate within the protected site and emerge as adults. In several species, the female wasp is wingless and antlike and will crawl from plant to plant in search of hosts.

Sand Wasps, Square-Headed Wasps, Aphid Wasps, and Beewolves (Crabronidae)

The Crabronidae are a large family of solitary hunters that range in size from small to large (5 to 40 mm). This family was formerly part of the family Sphecidae and includes the sand wasps, big-headed wasps, aphid wasps, and beewolves. Adult male and female wasps feed on pollen and nectar and are commonly found on flowers, but the larvae feed as predators. Mated female wasps construct a burrow in sandy soil, within hollow twigs, or in abandoned bee nests within wood or soil. After constructing or selecting a burrow, female wasps spend their days hunting for insects, which they collect and paralyze by stinging and injecting them with venom. The female will stuff her burrow with these insects and lay an egg on one of them, which will hatch and consume all of the provisioned prey. Crabronidae larvae will pupate within the burrow before emerging as adults.

Sand wasps are a large group of genera within the Crabronidae that construct burrows in sandy soil. They are solitary, but it's not uncommon for several females to construct burrows in a small area that has ideal soil

△ *Not all solitary stinging wasps build nests. Bethylidae find a caterpillar, paralyze it with a sting, and lay eggs within it. They leave the host and then return to this location to feed their developing offspring.*

conditions. Within a genera, sand wasps tend to specialize on a particular prey. One impressive example are the cicada killers, which prey upon cicadas (Cicadidae). Cicada killers are among the largest solitary stinging wasps found in the United States and measure 30 to 40 mm. The western cicada killer (*Sphecius grandis*) and Pacific cicada killer (*Sphecius convallis*) are found in the western United States. The Eastern cicada killer (*Sphecius speciosus*) is found throughout the eastern United States, and the Caribbean cicada killer (*Sphecius hogardii*) inhabits Florida.

Photo: Joseph R. Coelho

△ *A wasp that grabs cicadas in midair? Frighteningly, yes, wasps can be that big! Luckily cicada killers are not aggressive toward humans and are not social, so you will not find a massive paper nest of them. If you are lucky, however, you may see a female delivering a paralyzed cicada to her nest on the soil surface.*

Females often catch and paralyze their cicadas on the wing. The Eastern cicada killer also mates on the wing, with the wasps facing away from each other but both using their wings to remain aloft. Mated females construct nests in the soil with a number of cells, which they fill with multiple paralyzed cicadas. One egg will be laid within each cell. Females can determine the sex of the egg and will deposit female eggs within cells with the most abundant food. Although they are large and intimidating, female cicada killers rarely sting unless they are stepped on or handled. Males are often found in groups where they challenge each other for mates. It can be alarming to come across a group of these large wasps, but because males cannot sting, they do not pose a danger to humans.

As the name suggests, square-headed wasps have a quadrate head, but beyond this unique characteristic, many species look somewhat similar to some species in the Vespidae and Sphecidae families. They are often found in flowers, feeding on nectar and pollen. Many, such as those in the genera *Crabro* (8 to 12 mm) and *Ectemnius* (6 to 8 mm), are black with yellow stripes or

Photo: Beatriz Moisset

△ *This square-headed wasp in the genus* Ectemnius *is covered with pollen from feeding on flowers.*

markings on their abdomen, making identification in the garden difficult.

Aphid wasps (5 to 9 mm), which are typically slender and black, provision their nests with paralyzed aphids (Aphididae) and other soft-bodied plant feeders. Adults feed on aphid honeydew—a sugar-rich secretion aphids produce as they feed.

Beewolves (*Philanthus*, 7 to 9 mm), in the family Crabronidae, provision their nests with bumblebees, honey bees, or solitary bees, so they are not really beneficial for pest suppression. They do, however, contribute to overall backyard biodiversity.

Flower Wasps (Scoliidae)

Flower wasps—also called mammoth wasps, scarab hawks, or scarab hunters—are medium to large (½ to 1½ inches [1.5 to 4 cm]) and black with yellow or orange markings and wings with longitudinal wrinkles toward their tips. These wasps can be found throughout the United States, but are more diverse in the southern states. In residential landscapes, they can be found on grass lawns, where they search for the white grubs (Scarabaeidae) that feed on grass roots. They will dig into the soil to locate the grub, sting to paralyze it, and lay an egg on it. The larva develops by consuming the grub, pupates below ground, and then emerges from the lawn as an adult wasp.

Campsomeris is a common flower wasp genus in the United States. These species are ½ to 1¼ inches (1.5 to 3 cm) and are typically brown to black with yellow banding or spots on their abdomen. *Trielis octomaculata* looks similar to *Campsomeris* species and is found within the northern midwest, northeast, and southeastern United

△ *Aphid wasps feed on aphids and other plant feeding pests. This* Pseneo *species prefers sharpshooters in the family* Cicadellidae.

△ *Beewolves* (Philanthus) *are common in flowers where they hunt for prey and feed on pollen and nectar. Some species will prey on honey bees, but they are considered a minor pest in most cases by beekeepers.*

States. The genus *Scolia* takes its name from the curved body posture of the wasps that make up this genus. The blue-winged wasp (*Scolia dubia*) (¾ to 1 inch [2 to 2.5 cm]) is black, has a reddish-brown abdomen with two prominent yellow spots, and dark bluish wings. Males and females can be spotted flying in a figure-eight pattern just above a lawn surface in their courtship dance. Mated females will begin searching for grubs, including June beetles

The Flower Wasps (Scoliidae)

Scoliidae are not only lovely additions to the garden but useful as well. They prey on a number white grubs in the family Scarabaeidae, including June beetles and Japanese beetles. Female wasps dig into the soil to find a grub, which is paralyzed by the wasp who will then lay an egg onto the immobile grub. Her larva develops by consuming the grub belowground. It will then pupate and emerge from the soil as an adult.

(Campsomeris)
½ to 1¼ inches (1.5 to 3 cm)

Photo: Hannah Nendick-Mason

Double-Banded Scoliid
(Scolia bicincta)
¾ to 1 inch (2 to 2.5 cm)

Photo: Lynette Schimming

Blue Winged Wasp
(Scolia dubia)
¾ to 1 inch (2 to 2.5 cm)

Photo: Dan Leeder

(Scolia nobilitata)
½ to ¾ of an inch (1.5 to 2 cm)

Photo: Jason J. Dombroskie

(Triscolia ardens)
¾ to 1 inch (2 to 2.5 cm)

(Scarabaeidae: Melolonthinae) or Japanese beetles (Scarabaeidae: *Popillia japonica*). Other *Scolia* species include the double banded scoliid (*Scolia bicincta*) (¾ to 1 inch [2 to 2.5 cm]), which has a dark body with two cream bands on its abdomen and is found throughout the eastern and central states, and *Scolia nobilitata* (½ to ¾ of an inch [1.5 to 2 cm]), marked with four to six yellow to orange spots, which is found in the southeast. *Triscolia ardens* (¾ to 1 inch [2 to 2.5 cm]) is black with a red abdomen and can be found in the southwest and California.

Velvet Ants (Mutillidae)

Despite their common name, Mutillidae are not in the ant family (Formicidae). They take their name from the female's resemblance to ants. Velvet ants, however, are very different in that they are solitary. They are not common in the garden but may be found there, especially in drier regions of the United States. They range in size from small to large (3 to 23 mm), are dense with hair, and dark, often with bright orange, red, or yellow markings as seen in many *Dasymutilla* (½ to 1 inch [1.5 to 2.5 cm]) species. In a few species, such as the thistledown velvet ant (*Dasymutilla gloriosa*) (½ of an inch [1.5 cm]), females are covered with long, white hairs and resemble a thistle head. Male velvet ants are usually winged and look similar to many other wasps, but are densely hairy. They may be found feeding on the nectar of flowers. In some species, the larger male will carry the female in flight as they mate. Females are wingless and can be found on the soil surface, under rocks, and among grasses and vegetation. They are capable of making a squeaking sound when disturbed; consider this a warning not to handle a female velvet ant! They have a painful sting, which has resulted in common names such as cow killer (*Dasymutilla occidentalis*) (½ to ¾ of an inch [1.5 to 2 cm]), which is found in the southeastern United States. Females locate prey, including beetles (Coleoptera), caterpillars (Lepidoptera), flies (Diptera), bees and wasps (Hymenoptera), and even cockroaches (Blattoidae), and paralyze them with a sting. They often attack pupae or egg masses of other insects, and their larva will develop within them.

Photo: Mike Deep

△ Although it is unlikely that many cows have felt the sting of the cow killer velvet ant, this is the common name of the black and red Dasymutilla occidentalis. *The wingless females can be found primarily in sandy soil.*

Photo: Arlene Ripley

△ I have not had the opportunity to view the lovely thistledown velvet ant, Dasymutilla gloriosa, *in person. Seeing this thistle head mimic is yet another reason to plan a trip to the southwest!*

Photo: Roy Brown

△ Male velvet ants are winged and very hairy. If you want to handle a velvet ant, I advise sticking with winged ones; as with all wasps, the males do not sting.

Spider Wasps (Pompilidae)

Pompilidae are small to large (3 to 50 mm) in size and typically have dark bodies and wings and long spiny legs. The antennae of female spider wasps are curled, whereas the male's antennae are not. They may be all black, black with light markings, or metallic. Given the fact that they look very much like other solitary wasp species, spider wasps may be most easily identified if you find a female with its prey.

All spider wasps are soil nesting and provision their nests with spiders. Like other solitary wasps, female Pompilidae sting and paralyze their prey. Some species sting a spider in its burrow and then allow their larva to develop there, while the females of other species can be seen dragging a paralyzed spider to a burrow they constructed in sandy soil. Males and females can also be spotted on flowers feeding on pollen and nectar.

Among the larger spider wasps are the tarantula hawks (*Pepsis*) (½ to 2 inches [1.5 to 5 cm]), which are bluish-black wasps, often with orange wings, and sometimes orange legs and/or antennae. They occur mainly in the southwest, and adults are often seen feeding on milkweed plants. As the name indicates, females hunt tarantulas (Theraphosidae) by lurking around their burrows.

Photo: Edward Trammel

△ *A blue-black spider wasp (Anoplius) drags a paralyzed spider toward its burrow.*

Photo: John and Jane Balaban

△ *A female tarantula hawk (Pepsis) is able to paralyze a large tarantula, which will serve as food for her offspring.*

Thread-Waisted Wasps
(Sphecidae)

Wasps in the family Sphecidae range from ½ to 1¼ inches (1.5 to 3 cm) and are called the thread-waisted wasps due to their long, stalked abdomen. These wasps are mainly brown or black, and many have red, yellow, or metallic markings. Like the Crabronidae, most Sphecidae nest either below ground, construct nests in plant stems, or utilize abandoned bee nests. Some may also construct aerial nests of mud or grass. Adults are commonly found on flowers and feed on pollen and nectar. Some also feed on honeydew, the sugar-rich secretion produced by aphids, as well as the body fluids of the prey they collect for their larvae.

The cutworm wasps (*Podalonia*) (¾ to 1 inch [2 to 2.5 cm]) are a very beneficial group of thread-waisted wasps that can be found in home gardens. These wasps may be metallic, all black, or black with red markings. Females can be most easily identified if caught in the act of hunting cutworms (Noctuidae), which they feed to their offspring. Once the cutworm is located, stung, and paralyzed, the female will take it to a safe location while she constructs a soil burrow. After the burrow is completed, she will retrieve her cutworm and drag it into the burrow. The female will then lay one egg on the paralyzed prey and her offspring will consume it, pupate, and then emerge from the soil burrow as an adult.

You can guess what mud wasps and mud daubers use to construct their nests. These thread-waisted wasps feed on spiders, so they do not contribute to control of garden pests, but they are fascinating insects to watch in the garden. The black and yellow mud dauber (*Sceliphron*

Photo: Kim Moore

△ *A female cutworm wasp* (Podalonia) *struggles to drag a large cutworm toward her burrow.*

caementarium) is a large (1 to 1¼ inches [2.5 to 3 cm]) common species found throughout the United States. It has a black body with yellow markings and a very long "wasp waist" that is more than twice as long as the rest of its abdomen. Adults can be found feeding at flowers or collecting mud along ponds or puddles to make their nests. The nests consist of cylindrical cells of mud, and approximately forty trips are needed to complete one cell. The mud dauber wasp will also carry water in its mouth back to the nesting site to use for molding the mud. Nests are constructed on rock ledges, in cracks of tree bark, and on all sorts of structures and buildings. The nest can have up to twenty-five cells, each provisioned with several spiders. Between foraging trips, the wasps will place a thin cap of mud over each cell. Once they've collected enough spiders to sustain the developing larvae, females lay one egg in each cell and then seal it with mud.

The blue mud wasps (*Chalybion californicum* and *Chalybion zimmermanni*), which measure ½ to 1 inch (1.5 to 2.5 cm) and have a metallic blue body and wings, are into home restoration. They find abandoned mud nests of other wasps, including the black and yellow mud dauber,

△ *The black and yellow mud dauber (Sceliphron caementarium) will commonly build its nest on manmade structures, so check your garden shed eaves for this large wasp; perhaps you will get a chance to watch a female construct and provision her nest.*

△ *Blue mud wasps (Chalybion) are striking with metallic blue bodies and dark wings.*

and make it their own. Blue mud wasp females collect and carry water and use it to reshape the nest to fit their needs. Like mud daubers, these wasps also feed on spiders, including many web-building species. Female wasps will land on a spider web and mimic a trapped insect to draw the spider near so she can sting and paralyze it. This is not 100 percent successful though, and occasionally the wasps do become trapped within the web. Female blue mud wasps provision each cell of their nest with several spiders, lay one egg per cell, and seal them with mud. The larva will consume its spiders, pupate, and emerge as an adult by chewing an exit hole through the mud wall of the nest cell.

△ *Female blue mud wasps (Chalybion) are not afraid to commit to a fixer-upper. Instead of going to the trouble of constructing a new nest, they reconfigure one abandoned by another wasp. These images show a mud nest which has been reshaped by a blue mud wasp to protect her developing larvae. In the cross-section you can see two cells that have been provisioned with prey.*

Grass-carrying wasps (*Isodontia*) (¾ of an inch [2 cm]) are dark, thread-waisted wasps with reddish brown wings that are often found carrying grass to their nests. Females construct a nest in a tree cavity or other protected area, such as within window tracking or cracks or holes in window frames or siding. The nest consists of cells made of grass, which she provisions with crickets (Gryllidae) and katydids (Tettigoniidae). Depending on the species, each cell will include one or more larvae, which will feed on the stocked arthropod provisions. Like the grass-carrying wasps, several species within the genus *Sphex* also hunt grasshoppers (Acrididae), katydids (Tettigoniidae), and crickets (Gryllidae). Many are large, ranging from ½ to 1¼ inches (1.5 to 3 cm). The great golden digger wasp (*Sphex ichneumoneus*) has gold hairs on its head and pronotum, a plate covering the dorsal side of the thorax. It has a shiny orange and black abdomen and orange legs. The great black wasp (*Sphex pensylvanicus*) is shiny black and common in late summer in all but the northwestern United States. The katydid wasp (*Sphex nudus*) has a black body with gray markings on its head and light legs.

Yellow Jackets, Hornets, Paper Wasps, Mason Wasps, and Potter Wasps (Family: Vespidae)

Vespidae are small to large (5 to 25 mm) brown to black wasps that typically have yellow or orange markings. The distinctive feature for this group is that they fold their wings lengthwise when at rest. This family includes both solitary and social predatory wasp species.

△ *A female* Isodontia *constructs her nest within a hollow stem. Wasps and bees that nest in hollow stems can be encouraged by cutting back vegetation with this feature in the fall to provide nesting sites the following growing season. Also try inserting hollow sticks such as bamboo into the soil throughout the garden and watch for beneficial predator or pollinator colonizers!*

Photo: Patrick Coin

△ *The great golden digger wasp gets its name from the hairs covering much of its head and thorax. These wasps are common flower visitors. They prey primarily on grasshoppers (Acrididae), katydids (Tettigoniidae), and crickets (Gryllidae).*

Photo: Kevin Hall

△ *The great black wasp (*Sphex pensylvanicus*) is not only a predator, but also a pollinator like many flower-visiting wasps.*

Photo: Julie Feinstein

1

2

△ Odynerus *(1) and* Pterocheilus *(2) are both soil nesting mason wasps.*

Potter and mason wasps are medium-size (½ to ¾ of an inch [1.5 to 2 cm]) black wasps with yellow or white markings. These solitary wasps may construct their nests in pre-existing cavities such as holes in wood, in hollow stems, on the ground, on plants, or attached to structures. Mason and potter wasps use mud to construct their nests and they usually provision them with caterpillars (Lepidoptera). Wasps in several genera including *Odynerus* (½ to ¾ of an inch [1.5 to 2 cm]) and *Pterocheilus* (8 to 12 mm) construct their nests in the ground, and some species will build mud turrets over the nest entrance. Potter wasps *Eumenes* (½ to ¾ of an inch [1.5 to 2 cm]) and *Zeta* (¾ to 1 inch [2 to 2.5 cm]) have a long, slender pedicle. They get their name from their characteristic pot-shaped nest constructed of mud. Potter wasps transport mud in their mouth to a nesting site where they construct a roughly round or oval-shaped nest with several cells inside. The female does not enter the nest because it has too narrow of an opening, but she provisions each cell with several small caterpillars and then lays one egg per cell before covering each with mud.

△ *An* Eumenes fraternus *female constructed this nest on a flowering plant stem. She is carefully moving a caterpillar through a hole.*

△ *Once a potter wasp constructs its pot-shaped nest, the female can no longer fit inside. She will stuff the nest full of prey and then insert the tip of her abdomen to lay an egg onto these provisions. She then seals the nest with mud.*

Paper wasps are social insects that construct their nesting material by chewing wood and mixing it with saliva. They are medium to large (½ to 1 inch [1.5 to 2.5 cm]) and brown, typically with black and yellow markings. Males have yellow faces and curly antennae, while females have straight antennae. They construct paper combs that are exposed. In the spring, one or more females that mated the previous fall and overwintered will establish a new nest. The female begins a nest by attaching a thin paper stem to a substrate and creating a cell; then other cells are added with nests reaching 1,000 or more total cells. Females lay one egg within each cell, and the eggs hatch into all female larvae that they feed with bits of caterpillars the adults have collected. The adults will also increase the size of the paper cells as their offspring grow. These larvae will pupate within their cells and emerge as adults that look very similar to their mother in size and color, but they will not mate and instead will function as workers, tending to the growing colony. As these wasps take over tending the colony and foraging for food, the queen or queens that produced them will transition into only laying eggs, leaving the other duties to the workers. In mid-summer, reproductive males and females will produce eggs. Workers can lay unfertilized eggs, which result in male wasps, while queens can produce both reproductive males and females. These wasps will eventually mature to adulthood in late summer and leave the nest to mate. The mated females will overwinter either singly or in groups within protected areas such as tree holes, crevices in bark, leaf litter, old paper wasp nests, or even inside structures. The males, workers, and the current queens present within

△ *A female northern paper wasp* (Polistes fuscatus) *has just begun nest construction and has deposited an egg into each cell.*

the colony will die as winter sets in.

Polistes (½ to 1 inch [1.5 to 2.5 cm]) is a large genus of paper wasps with many species. The northern paper wasp (*Polistes fuscatus*) is common throughout much of the United States. This species is brown to black with yellow markings and dark antennae. The red paper wasp (*Polistes carolina*) is found in the southeastern and central United States and has a reddish-brown body and dark wings. The European paper wasp (*Polistes dominula*) was first discovered in the United States in the 1980s. It is often mistaken for a yellow jacket because of its shiny black body with yellow markings. But, unlike yellow jackets, the European paper wasp has orange antennae. There is concern that the European paper wasps may displace native paper wasp species.

Paper Wasps

Paper wasps are also sometimes called umbrella wasps because they create a nest that looks a tiny bit like an upside down umbrella. The "handle" of the umbrella is called a petiole (stalk) that anchors the nest, commonly to a soffit of a building or a tree branch. These paper nests are open: you can see the individual cells. This is what separates paper wasps from yellow jackets and hornets that cover their nest in a paper shell.

Photo: Patrick Coin

Mexican Honey Wasp
(Brachygastra mellifica)
6 to 9 mm

Photo: Jason D. Roberts

Red Paper Wasp
(Polistes carolina)
½ to 1 inch (1.5 to 2.5 cm)

Photo: Robin McLeod

European Paper Wasp
(Polistes dominula)
½ to 1 inch (1.5 to 2.5 cm)

Photo: Nolie Schneider

Northern Paper Wasp
(Polistes fuscatus)
½ to 1 inch (1.5 to 2.5 cm)

One of the most unique paper wasps is the Mexican honey wasp (*Brachygastra mellifica*) (6 to 9 mm). This black wasp with a yellow-striped abdomen produces and stores honey to feed to larvae. The workers collect nectar from flowering plants, and it is concentrated into honey and stored in cells at the edges of the nest. The honey is fed to larvae as their primary food source, but workers may collect insect prey as well. The Mexican honey wasp is found in southern Texas and Arizona.

1

2

△ *Nests made by hornets (1) tend to be more oblong or football-shaped, whereas those constructed by yellow jackets (2) are more rounded. Both hornets and yellow jackets cover their nest with a papery coating.*

Yellow jackets and hornets are medium to large (½ to 1¼ inches [1.5 to 3 cm]) black wasps with yellow or sometimes white markings. They are difficult to tell apart and can be best identified by the shape of their nest. Yellow jackets that nest aboveground tend to build more or less sphere-shaped nests, while hornets build nests that are more oblong and shaped like a football. Like paper wasps, yellow jackets and hornets use chewed wood mixed with saliva to construct their nest. They create a network of individual cells, but unlike paper wasps, this network is then encased in paper sheets. These nests may be constructed in holes underground or attached to branches or structures. Nests will have an opening at the base for wasps to enter and exit.

Like paper wasps, mated female queen yellow jackets and hornets overwinter and begin nest construction in the spring. Females will make a stalk that suspends the nest and then begin adding cells. An envelope of paper is constructed to cover the cells as they are added. The female will begin depositing her eggs within the cells, and the eggs will hatch and develop into female workers that do not reproduce. These workers will take over duties of nest construction and foraging, and the queen will stop all other tasks to exclusively focus on laying eggs. Larvae are fed nectar as well as small bits of fruit and parts of insects. These wasps hunt on the wing, where they keep watch for prey and then quickly dive down and grasp them. In mid-summer, reproductive males and new female queens are produced that leave the colony and mate. At the end of the season, the colony will begin to decline, and in northern regions with harsh winters, all but the new queens die off. In warmer parts of the United States, colonies of some species do not die off and may grow to include thousands of wasps.

There are three genera of yellow jackets and hornets. The genus *Vespa* includes only one species, the European Hornet (*Vespa crabro*) (½ to ¾ of an inch [1.5 to 2 cm], queens 25 to 35 mm), which is a non-native brown wasp with black and yellow markings. The abdomen of *Vespa crabro* has a significant amount of yellow present on the last several segments with small black markings. It commonly constructs its nest within tree holes or inside structures such as attics or garages. The genera *Dolichovespula* and *Vespula* include several species that look very similar, but have different nesting strategies. *Dolichovespula* tend to build nests above ground, whereas *Vespula* more commonly nest below ground. *Dolichovespula* includes the common aerial yellow jacket (*Dolichovespula arenaria*) (½ to ¾ of an inch [1.5 to 2 cm]) and the bald-faced hornet (*Dolichovespula maculata*) (1 to 1½ inches [2.5 to 4 cm]). Bald-faced hornets differ from yellow jackets in that they are more robust, and they have white markings on their head, thorax, and last few abdominal segments. The parasitic yellow jacket (*Dolichovespula arctica*) is an interesting member of this genus because it lays its eggs in the nests of other species, which raise its young. This species, then, does not produce workers.

There are several species of *Vespula* in the United States, including the common Eastern yellow jacket (*Vespula maculifrons*) (1¼ to 1½ inches [3 to 4 cm]), Western yellow jacket (*Vespula pensylvanica*) (1¼ to 1½ inches [3 to 4 cm]), Southern yellow jacket (*Vespula squamosa*) (¾ to 1 inch [2 to 2.5 cm]), blackjacket (*Vespula consobrina*) (½ to ¾ of an inch [1.5 to 2 cm]), and German yellow jacket (*Vespula germanica*) (½ to ¾ of an inch [1.5 to 2 cm]). The German yellow jacket is a non-native species that has become the dominant yellow jacket in many parts of the United States.

Yellow Jackets and Hornets

The family Vespidae includes the social yellow jackets and hornets. There are many species of these wasps found in the United States, and from the pictures below, you can see that they all look fairly similar: dark bodies with light yellow to white markings. These wasps all build round or oblong papery nests that workers provision with prey for developing larvae.

Photo: Libby & Rick Avis

Parasitic Yellow Jacket
(*Dolichovespula arctica*)
(15 to 18 mm)

Photo: Lynette Schimming

Common Aerial Yellow Jacket
(*Dolichovespula arenaria*)
½ to ¾ of an inch (1.5 to 2 cm)

Photo: Kevin Hall

Bald-Faced Hornet
(*Dolichovespula maculata*)
1 to 1½ inches (2.5 to 4 cm)

Photo: Patrick Coin

European Hornet
(*Vespa crabro*)
½ to ¾ of an inch (1.5 to 2 cm)

Photo: Limny & Rick Avis

Blackjacket
(*Vespula consobrina*)
½ to ¾ of an inch (1.5 to 2 cm)

Photo: Ilona L.

German Yellow Jacket
(*Vespula germanica*)
½ to ¾ of an inch (1.5 to 2 cm)

Photo: Andrew Meeds

Common Eastern Yellow Jacket
(*Vespula maculifrons*)
1¼ to 1½ inches (3 to 4 cm)

Photo: Matt Edmonds

Western Yellow Jacket
(*Vespula pensylvanica*)
1¼ to 1½ inches (3 to 4 cm)

Photo: Roy Cohutta Brown

Southern Yellow Jacket
(*Vespula squamosa*)
¾ to 1 inch (2 to 2.5 cm)

Ants (Formicidae)

Ants are social insects in the family Formicidae. There are hundreds of ant species within the United States, ranging in size from small to medium (2 to 20 mm) and from yellow or tan, to reddish, brown, or black. They nearly always have elbowed antennae and a hump-shaped second abdominal segment.

Nearly all ants are social insects that live in colonies of anywhere from a few dozen to more than a million individuals. The colony will consist of one or more queens that are responsible for egg laying. Eggs will hatch into legless larvae that are tended by wingless female workers that do not reproduce. The larvae are fed regurgitated food collected by the workers, including prey, nectar, and seeds. When it is time for a colony to reproduce, both males and females are hatched. These winged ants will leave the colony to mate. The females will shed their wings and attempt to establish a new colony or join an existing one.

In addition to tending to the larvae, workers maintain and defend the colony. As with wasps, some ants can sting and others cannot. Those that do not sting may produce a mixture of formic and other acids to repel enemies or kill prey. These ants have a pore at the tip of their abdomen from which they can eject this acid several centimeters!

In some species, there are many types of workers with specific duties. Soldier ants, for example, have larger heads and stronger mandibles that help them more effectively defend the colony. In fact, big-headed ants (*Pheidole*) were so named due to the large heads of the soldiers. Soldiers are 4 mm in length, whereas the

△ *A variety of insects store food for later consumption, either by themselves, their offspring, or other members of a colony. Some honey pot ant workers called repletes store this food within their own bodies. They eat a large amount until their abdomen swells and then store this food until a fellow ant needs it. Other workers will stroke the antennae of the honeypot ant, which causes it to regurgitate stored liquid that the worker can consume.*

smallest workers are half that size. Native big-headed ants are found in the southern United States along with an introduced invasive species (*Pheidole megacephala*), which is found in Florida and is believed to displace native species. The job of worker ants may also change with age. One unique example is the honeypot ant (*Myrmecocystus*), a genera in which some young workers are fed until they become "honey pots," ants that store food for the colony.

Ants can have both a positive and negative influence on garden pest management. Most species eat a mixed diet of arthropod prey, and large colonies are capable of removing a tremendous biomass of insects! It's important to note, however, that ants can also protect garden pests or become pests themselves. Many pest ants have been introduced into the United States accidently. A few of these, such as the red imported fire ant (*Solenopsis invicta*)

(3 to 6 mm) and the tawny crazy ant (*Nylanderia fulva*) (2 to 4 mm), are considered harmful invasive species.

The pavement ant (*Tetramorium caespitum*) (2 to 4 mm) is among the most common ants in home landscapes. They nest in soil and get their name from their propensity to excavate their nest through a crack in a sidewalk or driveway; small piles of soil are commonly seen at the nest entrance. This ant does feed on other insects, but they commonly invade homes where they will also feed on accessible food. The Argentine ant (*Linepithema humile*) (2 to 3 mm) is another common ant found in lawns and gardens that consumes other insects but also enters homes in search of food and water.

Carpenter ants (*Camponotus*) (4 to 13 mm) can become pest by constructing nests within the decaying moist or hollow wood in homes. The ants don't actually eat the wood—they feed on insects and other foods and are one group of ants that have developed a mutualistic association with aphids. These close associations between insects are known as myrmecophily. The ants protect the aphids from attack by predators and parasitoids, and in return, they collect the honeydew, the sugar-rich secretion

Photo: Arlo Pelegrin

△ *The red imported fire ant* (Solenopsis invicta) *is native to central America and was found in the United States in the 1930s. It is now present across much of the southern and western states. Colonies of this ant can reach several thousand workers, and the soil mound marking their colony entrance can be easily seen on a lawn. Take care to avoid or remove (using an insecticide) these mounds because fire ants will attack any perceived threat that comes in close contact to the entrance. Their sting is painful and causes a white pustule to form approximately twenty-four hours later.*

Photo: Graham Montgomery

△ *The tawny crazy ant* (Nylanderia fulva) *was detected in Texas in 2002, but they are also now found in Florida, Georgia, Louisiana, and Mississippi. This ant is native to South America and is named for its quick, erratic movements. These ants live in very large colonies and can become so abundant in yards that humans and their pets no longer want to spend time outside. This ant has also displaced many native and non-native ant species within its invaded range, including the red imported fire ant! Unfortunately, most people who have lived with both ants would take the fire ant over the crazy ant.*

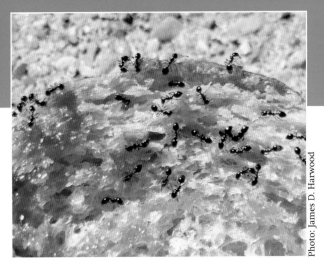

△ Pavement ants are a non-native species introduced from Europe but have been present in the United States for more than 200 years. They are dark brown to black, and they eat a variety of foods, including nearly anything that has fallen onto your kitchen floor. They are a common household pest.

△ Some ants develop close associations with other insects, a behavior known as myrmecophily. Several genera of ants "tend" aphids. They feed on honeydew, a sweet sap the aphids produce as they feed. To ensure a continued supply of this food resource, the ants protect the aphids from predators such as lady beetles (Coccinellidae). In the context of aphids, myrmecophily negatively affects any biological control the ants may provide.

produced by the aphids as they feed on plant sap. Ants may tend aphids that feed on roots below ground or those that feed on plant leaves and stems above ground. One ant will tend to many aphids, and sometimes, they will even move their aphids around on a plant or carry them between plants to ensure the aphids have an adequate and high-quality food source. Carpenter ants are not the only ants that tend aphids; many species and genera, including the little black ant (*Monomorium minimum*) (1 to 2 mm), cornfield ants (*Lasius*) (2 to 4 mm), and wood ants (*Formica*) (4 to 8 mm), also can have this mutualistic relationship with aphids. In cases where the other insect is a pest, such as carpenter ants and aphids, the ants' behavior interferes with pest suppression in gardens.

Ants can also tend insects that are not considered significant garden pests. Species in the genus *Liometopum* (3 to 6 mm) have a mutualistic relationship with butterfly larvae. Butterfly larvae in the family Lycaenidae feed on leaves, and caterpillars in this family have evolved to produce a sugar-rich sap to feed to ants. In exchange, the ants protect the larvae from predators and parasitoids, and their presence has shown to greatly enhance the larvae's chances of survival.

△ Ants can also protect species that are not seen as garden pests, such as the butterfly larvae in the family Lycaenidae, commonly known as the blues, coppers, hairstreaks, and harvesters. Some species in this family of butterflies have evolved to produce honeydew, which is fed to ants. The ants protect the larvae from predators, increasing their chance of survival to the pupal stage.

10

Spiders (Araneae)

Spiders are probably the most common predatory group in the garden, maybe more common than some gardeners would like. They also happen to be the most likely arthropod to be brought to my office dead (and sometimes squished), for identification. These predators are not aggressive, and only very few species in the United States can be a concern to humans. In fact, spiders are very beneficial hunters. They are primarily carnivores that eat a large diversity of arthropod prey in gardens, including many damaging pests. Spiders also feed on pollen and nectar and other plant material, so they benefit from the presence of these resources.

Spiders are in the order Araneae and have two major body segments: the cephalothorax and the abdomen. Most spiders have eight eyes on their cephalothorax, mouthparts called chelicerae, four pairs of legs, and pedipalps that look like a shorter fifth pair of legs. The pedipalps are used to manipulate prey and in courtship and mating. In males, the pedipalps are enlarged and serve as way for the male to transfer sperm to the female. The abdominal segment contains the spinnerets, which are the silk-producing organs, and some families also have a cribellum, another silk-spinning organ. Spiders use silk for many things, including trapping prey, lining burrows or retreats, protecting their eggs, dispersal, and courtship. Spiders exhibit incomplete metamorphosis. They hatch from eggs as tiny immatures called spiderlings and will undergo several molts, including later immature stages when they are referred to as juveniles or sub-adults (which have partially developed sexual organs), before reaching the adult stage.

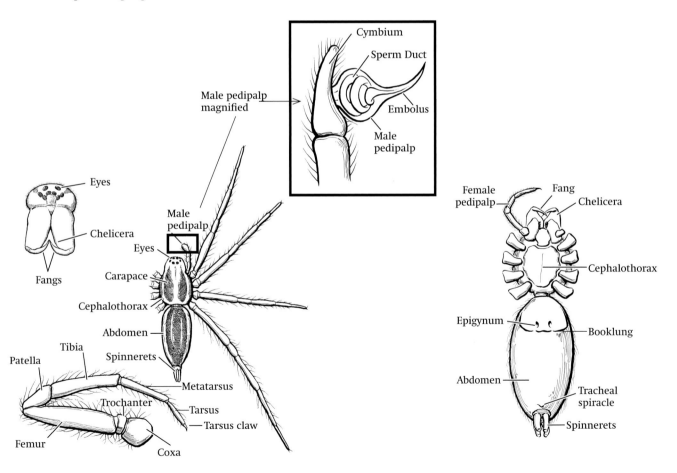

△ Here's a close-up look at a spider's basic body parts.

Once they reach adulthood, spiders have an interesting mating strategy. Males will deposit sperm from their genital opening onto a silk thread or small web and then take it up with their pedipalp. The male spider then stores the sperm until he can successfully court a female, when he will use his pedipalp to transfer the sperm to her genital opening.

Although spiders are all wingless predators, they are able to disperse effectively. Many are fast runners, yet to travel long distances, they can also rely on ballooning, a technique where spiders produce a stand of silk that catches the wind and allows the spider to "ride" the silk to a new location. They can travel several hundred miles this way and reach high altitudes. Most spiders balloon in early immature stages, though some can balloon throughout their entire life cycle.

Although they all produce silk, not all spiders use it to make webs that catch prey. This chapter discusses common families of active hunters that forage for prey, sit-and-wait families that wait for insects to come within their grasp, and several web-building families of spiders. Those that do build webs construct a variety of different types; the structure of the web often is so specific that you can identify the family of the builder without actually seeing the spider itself.

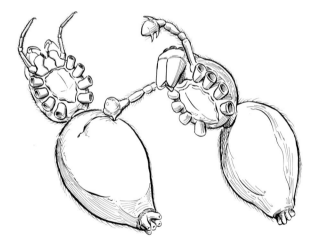

△ Spiders do not mate directly like insects do. Instead, a male spider will deposit sperm onto a thread of a web or a small web that he constructs specifically for this purpose. He will then use his pedipalp to take up the sperm from the web and store it. The male spider then uses his pedipalp to transfer his sperm to his mate.

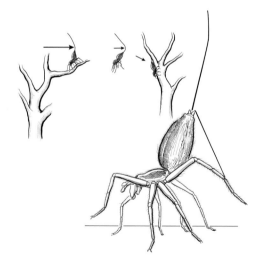

△ Spiders use silk in many interesting ways, even as a mode of dispersal! Spiders can move great distances by producing a silk thread to catch the wind, a process called ballooning.

Funnel-Web Weavers (Agelenidae)

unnel-web weavers are small- to medium-size (5 to 18 mm) spiders that are light gray or brown and often have a stripe running down their cephalothorax and abdomen, both of which are covered with hairs. The abdomen may also have chevron, or V-shaped patterns. One way to separate these spiders from wolf spiders, which look somewhat similar, is their long spinnerets. Two spinnerets can be easily seen on Agelenidae without the aid of a microscope, whereas on a wolf spider, they are shorter and more difficult to see. Common funnel-web weavers include grass spiders in the genus *Agelenopsis*, with different species found throughout the United States, and the hobo spider (*Eratigena agrestis*), which is a non-native species found in the Pacific Northwest.

If you find a funnel-web weaver in its web, identification is much easier. They construct a unique flat-sheet web for prey capture that has a narrow funnel on one end. The funnel provides a protected space for mating and keeps the egg sac safe as spiderlings develop. Webs may be attached to the sides of buildings or constructed in the grass or on shrubs. These spiders are nocturnal hunters, so typically you will find them in the funnel part of the web during the day. When a prey item lands within the sheet part of the web, the spider will dart out and bite it to paralyze it and then drag the prey back into the funnel to consume it.

▷ *Grass spiders in the genus* Agelenopsis *commonly construct their funnel-shaped web in home landscapes. These and other spiders in the family Agelenidae have long spinnerets, which help to distinguish them from wolf spiders (Lycosidae) when they are not in their web.*

Photo: Sarah Jane Rose

△ *The hobo spider (*Eratigena agrestis*) is a non-native species that was introduced into the Pacific Northwest in the 1930s.*

Photo: Sarah Jane Rose

△ *The family Agelenidae takes its common name funnel-web spiders from the characteristic funnel-shaped web these spiders construct. The spider is most commonly found inside the funnel, so its web may appear abandoned.*

Photo: Mary M. Gardiner

Orb Weavers (Araneidae)

The Orb weavers are among the largest spider families in terms of numbers of species. Their size ranges from small to large (2 to 30 mm), and they vary greatly in color, patterns, and body shape. These spiders have eight equally sized eyes; they rely more on web vibration than vision to detect prey trapped in their webs. In some species, males are much smaller than females and may vary in color as well. Their body color may range from drab and mottled brown, black, and gray to bright patterns that include white, green, red, and yellow. The abdomen of orb weavers is typically longer than it is wide and sometimes has spines or is uniquely shaped with pointed projections.

Most orb weavers spin what many think of as the classic spider web that looks like a wheel with many spokes. Some species also construct what are called web decorations, or stabilimenta, a crisscrossed weave of silk in the center of their web. When an insect becomes ensnared in the web, the spider will wrap it in silk and bite it to paralyze it. Many species build new webs frequently, sometimes daily, consuming the old web before finding a new site. Mating takes place within their web. In some species, a male will enter the web of a prospective mate and use his legs to pluck threads of the web, creating a vibration in an attempt to attract the female. Sometimes males vibrate a new thread they add to the web called a mating thread. After mating, females will spin a round egg sac in the fall, and spiderlings will emerge in the spring. In some species, one egg sac can contain hundreds of spiderlings.

Watching for Orb Weavers

Photo: Sarah Jane Rose

Black and Yellow Garden Spider
(*Argiope aurantia*)
Body length: male 5 to 6 mm, female ½ to 1 inch (1.5 to 2.5 cm)

Banded Garden Spider
(*Argiope trifasciata*)
Body length: male 4 to 6 mm, female ½ to 1 inch (1.5 to 2.5 cm)

Photo: John and Jane Balaban

Silver Garden Spider
(*Argiope argentata*)
Body length: male 3 to 5 mm, female ½ an inch (1.5 cm)

Shamrock Spider
(*Araneus trifolium*)
Body length: male 3 to 5 mm, female 6 to 10 mm

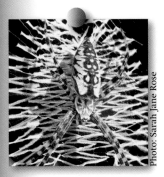

Photo: Sarah Jane Rose

Some orb weavers have been know construct stabilimenta. Several hypotheses have been proposed to explain the purpose of these structu including providing camouflage, making the spider appear larger, o making the web more visible to pre damage by birds or other animals.

There are many brightly colored and interesting shaped orb-weavers that inhabit home landscapes. They are also commonly found in tall grass and forested areas.

Marbled Orb Weaver
(*Araneus marmoreus*)
Body length: male 3 to 5 mm, female ½ to ¾ of an inch (1.5 to 2 cm)

Cat-Faced Spider
(*Araneus gemmoides*)
Body length: male 5.5 to 8 mm, female ½ to 1 inch (1.5 to 2.5 cm)

Cross Orb Weaver
(*Araneus diadematus*)
Body length: male 6 to 13 mm, female 6 to 20 mm

Spined Micrathena
(*Micrathena gracilis*)
Body length: male 3 to 5 mm, female 8 to 10 mm

Arrowshaped Micrathena
(*Micrathena sagittata*)
Body length: male 4 to 5 mm, female 8 to 9 mm

Spinybacked Orb Weaver
(*Gasteracantha cancriformis*)
Body length: male 1 to 3 mm, female 10 to 12 mm

Western Spotted Orb Weaver
(*Neoscona oaxacensis*)
Body length: male 5 to 12 mm, female ½ to ¾ of an inch (1.5 to 2 cm)

Arabesque Orb Weaver
(*Neoscona arabesca*)
Body length: male 5 to 6 mm, female 5 to 7 mm

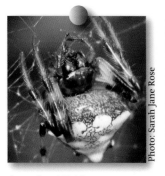

Arrowhead Spider
(*Verrucosa arenata*)
Body length: male 5 to 6 mm, female 5 to 7 mm

Starbellied Orb Weaver
(*Acanthepeira stellata*)
Body length: male 5 to 8 mm, female 7 to 15 mm

Humpbacked Orb Weaver
(*Eustala anastera*)
Body length: male 5 to 6 mm, female 5 to 8 mm

Sac Spiders (Clubionidae)

Sac spiders are small- to medium-size (3 to 10 mm) and pale yellow to brown, often lacking any patterning, although some species have chevron, or V-shaped, markings on their abdomen. They have eight equally sized eyes. Species in the genus *Clubiona* are found throughout the United States, and they are active at night when they wander through gardens hunting. These spiders do not build webs but use silk to construct saclike retreats on plant leaves, beneath loose bark, under rocks, or in the leaf litter. Females construct a silken sac to hold their eggs; these sacs are larger than retreats and are sometimes formed by rolling leaves together to create a protective cavity.

△ *Female sac spiders build a silken sac to protect their eggs. Sometimes, they may roll a leaf around the egg sac and tie it with silk.*

Two species of the long-legged sac spider, *Cheiracanthium inclusum* and *Cheiracanthium mildei*, are known to seek shelter inside houses in the fall when temperatures start to drop. They are not aggressive, but if bitten, the site should be monitored for possible swelling and slow healing of the wound. These species were formerly included in the family Clubionidae but are now placed within the prowling spider family Miturgidae.

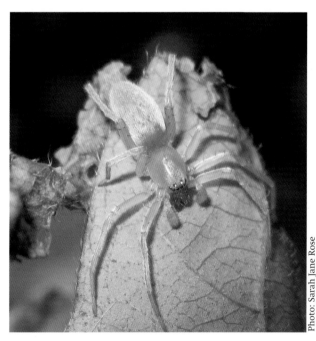

△ *Leaf-curling sac spiders (Clubiona) are yellow to brown and often lack any patterning on their body. These spiders construct retreats in plant leaves and other protected areas in the garden.*

△ *If you see a yellow to tan spider on the ceiling of your bathroom, chances are it's a long-legged sac spider (Cheiracanthium). These spiders can be found in gardens but commonly invade homes … or vehicles! Interestingly, long-legged sac spiders have found Mazda fuel line vents to be the perfect place to construct a retreat, causing damage to the fuel tank. A recall went into effect to remove the spiders and prevent them from reentering the lines.*

Mesh-Web Weavers (Dictynidae)

ictynidae are small to medium (1 to 8 mm) gray, brown, yellow, or reddish spiders that often have lateral banding on their carapace and may have chevron, or V-shaped, patterns on their abdomen. There are three subfamilies within this group with different foraging habitats and strategies. Members of the genera *Dictyna* construct irregular webs and attach them to grasses or branches and leaves of shrubs and trees to catch prey. Female *Dictyna* typically attach their egg sac to leaves. Some members of this genus have the rare habit of living communally within one web, with multiple males and females sometimes present. Spiders in the genera *Cicurina* and *Saltonia* construct their webs under rocks or logs or within the leaf litter and feed on soil and litter-dwelling arthropods. These spiders also place their egg sac outside of their web, typically attaching it below a rock or log.

Photo: Sarah Jane Rose

△ Female *Dictyna* *construct irregular mesh webs within vegetation. This female has constructed her web on a conifer branch.*

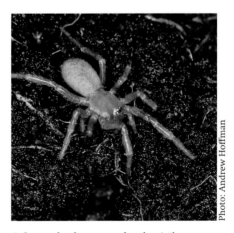

Photo: Andrew Hoffman

△ *Some mesh-web weavers, such as those in the genus* Cicurina, *construct webs under rocks, logs, or in leaf litter.*

Ground Spiders (Gnaphosidae)

The Gnaphosidae are small- to medium-size (2 to 17 mm) dark gray, brown, or black spiders that may have light colored spots, stripes, or chevron patterns on their abdomen. These spiders have stout legs, an oval abdomen, and barrel-shaped spinnerets. Common ground spiders include the eastern parson spider (*Herpyllus ecclesiasticus*), western parson spider (*Herpyllus propinquus*), and *Zelotes fratris*.

Ground spiders are active at night, when they hunt for prey along the soil surface, under rocks, and within mulch and leaf litter. These spiders spend the day in a silken retreat they construct for protection. In some species, females also guard their egg sac within this retreat.

Photo: Sarah Jane Rose

△ *The eastern parson spider (*Herpyllus ecclesiasticus*) has a barrel-shaped abdomen and thick stout legs typical of ground spiders.*

Sheetweb Spiders (Linyphiidae)

Sheetweb spiders are among the smallest garden spiders (1 to 7 mm). Other common names for this group include dwarf spiders or, in the United Kingdom, money spiders, a name from folklore that if one lands in your hair, it brings good fortune. These small spiders balloon throughout their lifecycle, so gardeners have a fairly good chance of having a sheetweb spider landing on them at some point.

Linyphiidae are smooth-bodied spiders with slender legs. Many, such as those in the subfamily Erigoninae, are very small and black, brown, or orange. However, many species in the subfamily Linyphiinae have a patterned abdomen such as the bowl and doily spider (*Frontinella communis*), filmy dome spider (*Neriene radiata*), and hammock spiders (*Pityohyphantes*).

Given the small size of these spiders, it is far easier to make your identification based on the web of this family, which are horizontal sheets of silk. Many species build small, flat sheet webs on or slightly above the soil surface, although some webs—such as those spun by the bowl and doily spiders—are more elaborate. These spiders use their webs to catch prey, including aphids (Aphididae), leafhoppers (Cicadellidae), springtails (Collembola), and small flies (Diptera). The spider will hang underneath the web and when a suitable prey lands within the silk, will run across the web and bite it from underneath, causing paralysis. Mating occurs in the web, and in some species, the male will destroy most of the web prior to mating to remove the female sex pheromone embedded in the silk to prevent the attraction of other males. Female sheetweb spiders produce a round egg mass near their web.

△ *Sheetweb spiders in the subfamily Erigoninae build small webs on the soil surface or just above it.*

△ *Here are a male and female filmy dome spider,* Neriene radiata.

△ *Here is the bowl and doily spider,* Frontinella communis.

△ *Here is a hammock spider,* Pityohyphantes.

△ *One look at its web and it is clear how the bowl and doily spider (*Frontinella communis*) got its name.*

Wolf Spiders (Lycosidae)

△ Wolf spiders (Lycosidae) have two large central eyes surrounded by six smaller ones arranged in three rows.

Wolf spiders are small- to large-size arachnids (3 to 35 mm) that hunt along the soil surface. Their gray, brown, or black coloration blends well into a background of mulch and soil. They often have a stripe running down their cephalothorax, which sometimes continues down their abdomen, and they are covered in dense hairs. Their abdomen may also have spots or chevron patterns. Wolf spiders have robust legs that sometimes have alternating light and dark banding, and they are able to run quickly across the garden floor. They hunt by sight and have eight eyes arranged in three rows—four and two small eyes are located on the top and bottom row, respectively, and in between these they have two large eyes that you can see without the aid of a microscope if you are willing to get close enough. They

may be active hunting prey during the day or at night, and despite their name, they are solitary hunters and do not hunt in packs. There are many common wolf spider species found in garden habitats, including the thin-

△ A female wolf spider seeks shelter in her burrow.

△ A thin-legged wolf spider (Pardosa) floats in water. Some wolf spiders are able to move throughout marshy habitats by traversing water in this way. Some may also "fish" for insects by hanging one leg below the surface of the water!

legged wolf spiders (*Pardosa*) and pirate wolf spiders (*Pirata*), and species in the genera *Hogna*, *Schizocosa*, and *Rabidosa*. These spiders do not build webs, but can use silk to line retreats and protect their developing eggs.

Male wolf spiders may wave their pedipalps or legs to court females, or they may also drum them on a substrate when in sight of a prospective mate. Mated female wolf spiders can be especially easy to identify in the garden because they carry their eggs with them. Females construct a round silken sac to carry their eggs and attach it to their spinnerets at the end of the abdomen. The female will carry her eggs with her until the spiderlings are ready to hatch. Her parental investment does not end there; after the spiderlings emerge from their egg case, they climb up their mother's legs and ride on her abdomen until their first molt.

△ *Here is a wolf spider in the genus* Rabidosa, *with its robust legs visible.*

△ *Here is a female wolf spider with her egg sac. The female will carry her eggs with her until they hatch.*

△ *Here is a wolf spider in the genus* Hogna, *blending in well with sand and gravel.*

△ *Newly hatched spiderlings cling to the abdomen of their mother. This female will carry her offspring until their first molt.*

Spiders (Araneae) **139**

Lynx Spiders (Oxyopidae)

Lynx spiders are small to large spiders (4 to 21 mm) and have an elongated, pointed abdomen, legs with spines, and a unique hexagonal eye pattern. These spiders range widely in color, but often have a light body with darker patterning, and some species have remarkable iridescent purple or gold markings. Species in the genera *Oxyopes* and *Hamataliwa* are small- to medium-sized spiders that are gray, brown, or yellow. The green lynx spiders, *Peucetia* spp., are larger, and as the name suggests, are bright green.

Oxyopidae are well camouflaged within the vegetation in which they forage; *Oxyopes* and *Hamataliwa* forage commonly on shrubs and trees, whereas *Peucetia* prefer tall grasses and herbaceous flowering plants. All Oxyopidae are ambush hunters; they may sit and wait for prey to come within reach or actively stalk their prey. These spiders do not build webs to capture prey, but female

Photo: Jeff Hollenbeck

△ *Here is a gorgeous green lynx spider* (Peucetia), *in the process of molting.*

Peucetia hang their egg sac within a mesh of silk and *Oxyopes* and *Hamataliwa* attach their egg sac to plant leaves and stems.

Photo: Sarah Jane Rose

△ *Here is lynx spider in the genus* Oxyopes; *note the elongated, pointed abdomen and spiny legs.*

Photo: Jeff Hollenbeck

△ *A female* Oxyopes *has just deposited her eggs. She will hold them against the twig to harden and then spin silk over them. The female will then guard the egg sac.*

Nursery Web or Fishing Spiders (Pisauridae)

Nursery web or fishing spiders are small to large spiders (1 to 37 mm) with long robust legs. Body coloration may be reddish, brown, or tan, often with a lateral strip running down their cephalothorax and abdomen. These spiders may also have pairs of light-colored spots or dark chevron patterns present on the abdomen.

Pisauridae can be found in a number of different habitats. Those in the genus *Dolomedes* are found in semi-aquatic habitats, where they hunt aquatic insects and earn the name fishing spiders because some species are known to catch small fish as well! These spiders also have the ability to walk on the water surface, so you may see them out in a garden pond or foraging around the water's edge.

Based on body size, shape, and coloration, Pisauridae may be confused with wolf spiders (Lycosidae). One way to differentiate between the two is to look at the eyes — Pisauridae have eight eyes of relatively equal size, whereas Lycosidae have two much larger central eyes. Like wolf spiders, females do carry their egg sac with them, but they carry it with their pedipalps and chelicerae instead of from their spinnerets. The other common name, nursery web spider, is derived from the female behavior of spinning a "nursery web," where she will suspend the egg sac when it is close to hatching. The spiderlings will hatch and stay in this nursery web until their first molt, when they move on to hunt on their own. The female keeps a close watch on the nursery web from nearby rocks or vegetation, where she is ready defend it if necessary. Some Pisauridae also have interesting mating rituals. Males in the genus *Pisaura*, for example, bring the female a fly wrapped in silk to feed on as they mate.

Photo: Sarah Jane Rose

△ Pisauridae can be confused with wolf spiders. Unlike wolf spiders, which have two large central eyes and six small ones, these spiders have eight eyes of equal size.

Photo: Matt Edmonds

△ Pisauridae females carry their egg sac with them using their mouthparts.

Jumping Spiders (Salticidae)

△ *The bold jumping spider* (Phidippus audax) *has a dark body with light spots on its abdomen.*

Photo: Sarah Jane Rose

There are many species of jumping spiders within the family Salticidae, ranging in size from small to large (1 to 22 mm). The body shape and behavior of this family make identification of jumping spiders a little easier than some others. They have a flattened body with a square-shaped cephalothorax and short, stout legs. Jumping spiders have eight eyes, two of which are very large; these spiders can see color and are visual hunters. They also vary widely in color and patterning. Most often, their bodies are dark with light or iridescent markings such as the zebra spider (*Salticus scenicus*) and the bold jumping spider (*Phidippus audax*), which are both very common jumping spiders found in residential landscapes. Other species, such as those in the genera *Myrmarachne*, *Peckhamia*, and *Synageles*, mimic ants, some are shiny and beetlelike (*Metacyrba*), while others have mottled patterns of brown, black, gray, and white that blend well with vegetation, bark, or leaf litter (*Naphrys*, *Platycryptus*, and *Sitticus*). Those in the genus *Lyssomanes* are translucent and green.

In many species, males and females have different coloration. Males may have brightly colored pedipalps, legs, and abdomen because these are the body parts that they raise and shake during elaborate courtship dances. Some males also use their pedipalps to produce a vibration that females are able to detect.

Jumping spiders get their name from their ability to pounce forward and grasp potential prey. Most do not build webs and are day-active hunters. These spiders stalk their prey for some time, and, when within jumping range, they attach a strand of silk to the substrate to serve as a safety line. These spiders can jump up to twenty-five times their body length. This family also uses silk to construct retreats in which to avoid bad weather or spend the night. Females guard their eggs and spiderlings until they have molted and dispersed.

Photo: Ryan Kaldari

△ *The zebra jumper* (Salticus scenicus) *is a common spider in home landscapes.*

Photo: Sarah Jane Rose

◁ *A close look is required to see that this is in fact a spider and not an ant. The four pairs of legs and lack of antennae give this spider away.*

Brown Recluse Spiders (Sicariidae)

Photo: Christy Beal

△ *The translucent green jumping spiders* (Lyssomanes) *blend in so well to their surroundings that they can hide in plain sight on vegetation.*

These spiders are not necessarily common in the garden, but they are a spider group of concern due to their cytotoxic venom that can produce lesions and persistent sores in humans. Brown recluse spiders (*Loxosceles*) are medium-size (5 to 13 mm) with six eyes arranged in three groups of two. They are yellowish to brown and can have dark markings on their cephalothorax that resembles a violin with its neck pointed toward the abdomen of the spider. Brown recluse spiders build irregular webs in protected spaces such as wood piles and backyard sheds. Females will attach their egg sac to this webbing.

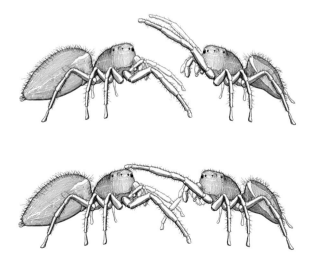

△ *Jumping spiders can have complex mating displays. Males often look much different from females and may have bright tufts of hair on their legs and/or colorful patterns on their abdomen. A male will perform a courtship dance for the female that may include raising and vibrating his legs and pedipalps, moving in an erratic zigzag fashion, and shaking his abdomen. Eventually, he will move forward and touch the female with his legs. If she is receptive to his advances and not aggressive, he will climb onto her and inseminate her with his pedipalp.*

Photo: Matt Edmonds

△ *Here is a brown recluse spider* (Loxosceles). *Note the fiddle-shaped marking on the cephalothorax.*

Long-Jawed Orb Weavers (Tetragnathidae)

Long-jawed orb weavers are small to large spiders (2 to 25 mm). As the name suggests, spiders in this family (especially males) have large mouthparts that protrude forward. They may be green, red, silver, or yellow. Many of these orb weavers, such as those in the genus *Tetragnatha*, have an elongated abdomen and long, thin legs. Their first set of legs is the longest, and the third pair of legs is noticeably shorter than the others. Some, however, are more stout bodied with a more rounded abdomen such those in the genus *Pachygnatha* and the small *Glenognatha foxi* (2 mm long).

Long-jawed orb weavers construct an orb-shaped web similar to that built by Araneidae. The spider sits in the hub of the web with the front two pairs of legs held forward and the back two held behind its body, looking like a twig or blade of grass. When the web intercepts potential prey, the spider leaps forward to capture the food item.

Photo: Sarah Jane Rose

△ It is not at all surprising that most species in the long-jawed orb weaver family, such as this *Tetragnatha* species, have long jaws!

Photo: Sarah Jane Rose

△ Some long-jawed orb weavers, such as those in the genus *Pachygnatha*, have a more rounded abdomen.

Cobweb Weavers (Theridiidae)

The cobweb weavers are a large and diverse family of spiders, ranging in size from small- to medium-size (1 to 12 mm). These spiders range in color from brown and black to bright green and red. They are also commonly referred to as comb-footed spiders because they have a row of spines on their back legs used to comb silk from their spinnerets. They typically have a very round ball-like abdomen and long thin legs, but some species do have a more elongated abdomen.

This family includes the black widow spiders (*Latrodectus*). The western (*Latrodectus hesperus*), southern (*Latrodectus mactans*), and northern (*Latrodectus variolus*) black widows are all found in the United States. These spiders commonly build their webs in protected places such as woodpiles or garden sheds. Females are dark black, often with a red or orange hourglass on the underside on their abdomen, although these bright markings may also look like two spots or be absent altogether. Males are smaller than females and are gray to brown. Bites to humans occur when the spider is accidently touched and can cause lesions, persistent sores, and in very rare cases, death from exposure to black widow spider venom.

Photo: Stephen Luk

△ The tiny long-jawed orb weaver Glenognatha foxi *looks very similar to sheetweb spider (Linyphiidae). It has a bright orange abdomen with light and dark patterning.*

Photo: Sarah Jane Rose

△ *Long-jawed orb weavers often rest with their legs outstretched in front of and behind them, making them look like a stick or blade of grass.*

Like most web builders, finding a cobweb weaver in its web greatly eases identification. These spiders build an irregular web of sticky silk. When an insect touches a web strand, the strand breaks, entangling the prey while the spider hangs upside down in its web to await these captures. Most cobweb weavers will spin their own webs, however, those in the genus *Rhomphaea* and *Argyrodes* forage in the webs of other spiders. These spiders feed on prey within the web and may also consume the other spider's eggs or the adult spider itself.

△ *Female black widow spiders (*Latrodectus*) are black with a red hourglass shape on their abdomen. In some species, such as the northern black widow (*Latrodectus variolus*), this pattern looks more like two red spots than a completed hourglass.*

△ *Most cobweb weavers have an orb-shaped abdomen and are often colorful.*

▷ *Spiders in the genus* Rhomphaea *make things easier on themselves; instead of constructing their own web, they forage for prey in the webs of other spiders.*

Crab Spiders (Thomisidae)

Named for their crablike sideways movement, Thomisidae are small- to medium-size (2 to 11 mm) spiders that have a flattened stout body. Their front two pairs of legs are long and robust, and the back two are shorter and more slender. Although they can produce silk for drop lines and to protect their eggs, crab spiders do not build webs. They are sit-and-wait ambush predators that hunt with the front two pairs of legs outstretched.

The color of a crab spider indicates where it hunts in the garden; those that hunt under loose bark such as the genus *Coriarachne* and those found foraging across mulch and leaf litter such as *Xysticus* are typically brown, black, or rust colored. Those species hunting in flowers may be bright white, yellow, green, or red. Some are even able to change color gradually to match the flower they are sitting on! Crab spiders consume a variety of arthropod prey, but those that inhabit flowers, such as *Misumena* and *Misumenops*, are known to commonly hunt pollinating insects such as bees and hover flies.

In some crab spiders, males cover their mate loosely with strands of silk prior to mating, with what is called a bridal veil. The females can easily escape the silk after mating, so this does not serve as a means to restrict her movement but is part of their courtship. Females construct a nest to house their egg sac, sometimes by rolling a leaf into a protective structure using silk. Females will guard this nest until the spiderlings hatch.

Photo: Sarah Jane Rose

△ Crab spiders often blend in well with their hunting grounds. Those that forage on leaf litter or on bark are often mottled brown, black, or rust colored.

Photo: Sarah Jane Rose

△ Can you spot the crab spider? Species that hunt in flowers are often brightly colored. Some can even change color to match their selected flower!

11

Other Arachnids (Arachnida)

Beyond spiders, there are a few additional predatory arachnids that are simply too remarkable to leave out of this book. Many of these odd creatures are legendary and have been the focus of strange and incorrect urban legends. Here you will learn their real story, which is fascinating in its own right.

In addition to spiders, the class Arachnida also includes mites, scorpions, and many others. Like spiders, the arthropods in this class have two body segments (cephalothorax and abdomen), four pairs of legs, pedipalps, and mouthparts called chelicerae. They also lack antennae and wings. These arthropods are generalists that feed on a vast array of prey, with both pests and beneficial species comprising part of their diet.

These arachnids develop via gradual metamorphosis, meaning they hatch from an egg (or in some cases, females give live birth) as immature nymphs and undergo several molts to reach adulthood with no pupal stage.

Some of these arachnids, such as mites (Acari) and harvestmen (Opiliones), are very common in home gardens. Others, such as whip spiders (Amblypygi) and wind scorpions (Solifugae), are rare but might be found by those willing to spend time in their garden after dark, particularly in the southwestern United States.

Predatory Mites (Acari)

The Acari includes mites and ticks. This mega-diverse order includes species that feed on plants and others that consume fungi, decomposing organic matter, and algae. Many species are parasites that attack invertebrates, while others parasitize vertebrates (such as ticks, chiggers, and scabies mites), and still others live as nonthreatening species with other animals, including the mites that live in the pores on all of our faces (*Demodex*)!

Here, we focus on families that include species of beneficial predators of common garden pests, including spider mites (Tetranychidae), thrips (Thripidae), whiteflies (Aleyrodidae), and other tiny plant feeders. The life cycle of predatory mites consists of an egg stage, larval stage, nymphal stages, and an adult stage. Females lay clear eggs that hatch into larvae. The larvae then molt through two additional immature stages called the protonymph and deutonymph stages (some other mites have a third nymphal stage—tritonymph). As larvae, predatory mites have six legs, whereas both nymphal and adult stages have four pairs of legs. Several generations of these predators can occur within one growing season, and complete development typically occurs within one to three weeks. The color of predatory mites varies from light yellow to dark red and can depend on their diet.

Perhaps the most important mites from a gardening perspective are the Phytoseiidae, a family that contains more than 2,000 species of minute (0.2 to 0.6 mm), voracious predators. Phytoseiidae can consume several adult spider mites (Tetranychidae) or dozens of eggs per day, making them major players in controlling pest populations.

Phytoseiidae feed on prey throughout their entire lifecycle, beginning in their larval stage. Their chelicerae allow them to pierce their prey and ingest its liquid contents. Some species feed on pollen or nectar as well, particularly when pests are uncommon. Some common Phytoseiidae found in gardens and home landscapes include *Neoseiulus californicus*, *Neoseiulus fallacis*, *Galendromus occidentalis*, *Typhlodromus pyri*, and *Phytoseiulus persimilis*. These mites feed on pests infesting fruit trees, vegetable crops, and ornamental plants. Depending on the climatic conditions, certain species will be more dominant; for example, *Typhlodromus pyri* is more effective in cool, humid climes, whereas *Galendromus occidentalis* prefers hot, dry climates.

Other mite predators include several dark red species common in home gardens—whirligig mites, concrete mites, and velvet mites. Whirligig mites (Anystidae: *Anystis*) range from 1 to 2 mm in body length and are generally orange to red. They have a truncated posterior and tapering legs that seem to radiate from a central point in the middle of their body. They are quite noticeable due to their bright color and incredibly fast, seemingly erratic running behavior that only ceases when they stop for a meal. They are voracious and have been known to eat a wide range of prey, including proportionately larger insects such as treehopper nymphs (Membracidae).

Velvet mites (Trombidiidae) are often found in or on the soil but climb plants in search of food during periods of high humidity, such as after a rain. Most species are between 1 and 8 mm, but some are the largest of all mites, reaching over 17 mm! The larval stage feeds as a parasite of many arthropods, including aphids

Photo: Peter Cristofono

△ *Watch for red whirligig mites (Anystidae: Anystis) scurrying quickly on plant leaves and stems in a quest for prey.*

(Aphididae), arachnids (including spiders and others), beetles (Coleoptera), flies (Diptera), and grasshoppers (Acrididae), and have been found to influence host populations by reducing fecundity and longevity. To find a host, the larval mites usually climb upward on plants and attach to a host by secreting a kind of glue with their mouthparts. When they leave their host, they develop into nymphs and finally adults, both of which forage as free-living predators of other arthropods and their eggs. In fact, some velvet mites are dominant predators of the same animal that they parasitize as larvae.

Concrete mites (*Balaustium*) in the family Erythraeidae are close relatives of the velvet mites and are frequently encountered in the garden. These mites forage for eggs, larvae, and pupae of pests throughout their lifecycle. Interestingly, they are also able to develop exclusively on pollen, which is a rare habit among arachnids. They are often found in great numbers on flowers, and providing these predators with pollen-producing plants will sustain populations when pests are scarce.

Photo: Rocco Saya

△ *Velvet mites (Trombidiidae) are parasites as larvae; this individual is feeding on a crab spider. First, the mite uses its legs to grasp onto the passing spider. Then, there is the matter of feeding on it. The mite secretes a strawlike structure (called a stylostome) that dissolves through the exoskeleton or "skin" of the spider, enabling it to suck up internal fluids.*

Many other families of mites include predatory species that may be found in gardens settings, but because many of these live in the soil, they often go unnoticed. Predatory species of Parasitidae, Bdellidae (snout mites), Cunaxidae, and Microtrombidiidae feed on many soft-bodied pests such as nematodes and fly larvae that are often garden pests. Some families well known for their plant-feeding pest species, such as Stigmaeidae and Tydeidae, also include other species that are effective predators.

▷ *Adult velvet mites (Trombidiidae) in the genus* Allothrombium *are effective natural enemies of aphids because they parasitize aphids as larvae and predate them as adults.*

Photo: Ray Fisher

The Lesser-Known Predatory Mites

Your backyard is teeming with mites. Don't start itching! We're talking about beneficial predatory mites here. These tiny hunters can be found foraging on plants within the soil and garden mulch. They feed on pest mites such as spider mites (Tetranychidae), along with other tiny arthropods and insect eggs. These mites are all less than 1 mm in length.

Pergamasus mite
(Parasitidae)

Snout mites
(Bdellidae)

(Cunaxidae)

Photo: Dan Leeder

Photo: Brian Valentine

Photo: Stephen Luk

Stigmatid mites
(Stigmaeidae)

Tydeid mites
(Tydeidae)

(Microtrombidiidae)

Photo: Arlo Pelegrin

Photo: Scott Justis

Photo: Nolie Schneider

Photo: Michael Charters

Although concrete mites (Balaustium) are predatory, in times of low prey availability, they can survive on a diet of pollen alone. To sustain a population of these mites at the ready to defend your plants from pests, provide them with a steady supply of flowering plant resources.

Whip Spiders (Amblypygi)

Whip spiders may look ferocious, but these 5 to 38 mm arachnids do not bite or sting humans or produce venom. They can, however, pinch with their pedipalps when picked up! Whip spiders have very flattened bodies with long raptorial pedipalps used for grasping prey, fighting, and in courtship and mating. These arthropods have eight eyes, but relatively poor vision. They rely on their first pair of thin antenniform legs to detect stimuli such as odors and vibrations from the surrounding environment. Whip spiders are ambush hunters that use their large pedipalps to quickly grasp unsuspecting arthropod prey.

Many arachnids have an elaborate courtship, and with whip spiders, this can last several hours. When courting a female, a male will first approach her and shake his antenniform legs. He will then move toward the female and grasp her pedipalp to initiate a "courtship dance." As the dance progresses, the male shakes his body up and down and taps his antenniform legs on the female's legs and pedipalps, as well as on the ground surface. Like many arachnids, the courtship dance concludes with the male depositing a spermatophore, or capsule containing sperm, on the ground. Then the male will guide the female over the spermatophore and tap repeatedly on her abdomen, pressing it down over the spermatophore until it is taken up by the female to fertilize her eggs.

When the female is ready to lay eggs, she finds a vertical surface, and with her head pointed upward, she will lay her eggs inside a viscous fluid that hardens the eggs and adheres them to the underside of her abdomen. When the immature whip spiders hatch, the young cling to their mother's abdomen for one to two weeks, during which time, the female does not move. After their first molt, they leave their mother and begin to forage on their own. Whip spiders undergo three to eight molts to reach the adult stage.

Photo: Richard Bradley

△ *Whip spiders (Amblypygi) range from 5 to 38 mm in the United States, but larger species are found in other parts of the world such as this specimen collected near the La Selva Biological Station in Costa Rica by Ohio State University graduate student Sarah Jane Rose. It was released shortly after this photo shoot. Although they look ferocious, Amblypygi do not bite or sting humans.*

Harvestmen (Opiliones)

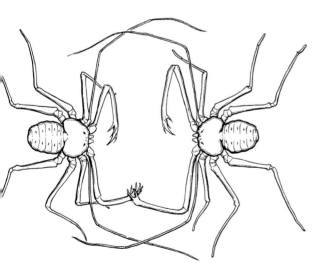

Photo: Martin Hauser

△ *This female whip spider has her newly hatched young with her. During the first one to two weeks after her eggs hatch, the female will not move. She waits until these offspring molt and leave her to forage on their own.*

△ *This male whip spider (Amblypygi) is shown here with a prospective mate. A male will first approach a female and shake his antenniform legs. He will then move toward her and grasp her pedipalp to initiate a courtship dance, which can last for several hours.*

Harvestmen (also called harvest spiders, shepherd's spiders, reapers, and daddy long legs) in the order Opiliones are referred to by several common names, some of which originate from folklore associating a high abundance of these predators during the harvest season. The name shepherd's spider originates in Europe, where the arachnid's body atop long legs was reminiscent of shepherds using stilts to observe their flocks. In the United States, the common name daddy long legs is most frequently used and describes the look of most species in this order, whose bodies range from 2 to 10 mm in length with legs several times this length. They can vary in color, but the majority are brown or grey and blend into soil, leaf litter, or mulch.

Like all arachnids, harvestmen have a two-segmented body with a cephalothorax and abdomen, but unlike other species in the class Arachnida, it does not appear this way because the segments do not narrow where they join, giving the appearance of a single-segmented oval body. All harvestmen have one pair of pedipalps, which are short appendages near their mouthparts used in feeding and mating and four pairs of walking legs. In some families, the pedipalps are enlarged with claws that are useful in prey capture. Although harvestmen look a lot like spiders, there are some important differences. First, harvestmen do not possess silk glands and cannot build webs. And, whereas spiders have several eyes, Opiliones have two and are only capable of distinguishing between varying intensities of light. Given their limited vision, other senses are very important, and their second pair of legs has receptors that can detect chemical cues from their environment. Opiliones also differ from spiders in that they do not use

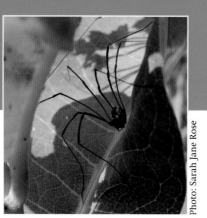

▷ *Phalangium opilio (Phalangiidae) is one of the most common harvestmen found in home landscapes.*

Photo: Sarah Jane Rose

Photo: Rodrigo H. Willemart

△ *Some harvestmen form aggregations, perhaps to protect the arachnids from harsh conditions and make it easier to find mates and avoid predators.*

venom for prey capture. Instead, they simply grasp prey with their chelicerae and pedipalps, tear it apart, and consume it. The urban legend that harvestmen have the deadliest venom but cannot penetrate the skin of humans to inject it is completely false!

These arachnids can be found foraging within the mulch layer of gardens as well as on vegetation, and they consume a vast array of arthropod prey. Most hunt using a sit-and-wait strategy, and most are active at night. They are unique among arachnids in that they will also consume dead prey (both invertebrates and vertebrates) and plant material and have even been found feeding on bird droppings.

Harvestmen mate directly, unlike many arachnids where sperm is transferred indirectly, either via a pedipalp in the case of spiders or as a spermatophore deposited on substrate as in many other groups. In many species, males compete for mates, and once they successfully court and mate with a female, they will often guard her until she lays her eggs. Some males also offer the female a "nuptial gift," a source of food produced from his chelicerae that is consumed during mating. Females have a long ovipositor that allows them to lay their eggs under bark and in the soil and leaf litter. Some species protect their eggs by covering them with debris. Females, and even males of some species, may also guard their eggs until they hatch.

One of the most common harvestmen in the garden is a species of European origin, *Phalangium opilio* in the family Phalangiidae. This species is widely distributed around the world, including in the United States, and is known to feed on aphids (Aphididae), caterpillars (Lepidoptera), leafhoppers (Cicadellidae), beetle larvae (Coleoptera), spider mites (Tetranychidae), and slugs.

Another family found within the United States is the Sclerosomatidae, comprising approximately 1,300 species, including the desert harvestmen (*Eurybunus*), dalquestia harvestmen (*Dalquestia*), and *Leiobunum*. Some harvestmen, including several species in the genus *Leiobunum*, are known to form daytime aggregations, possibly an adaptation to dealing with harsh temperatures or dry conditions. Aggregation may also be an adaption to improve the chances of finding a mate or avoid being eaten by a predator. Harvestmen are able to produce defensive secretions, and these may be more effective against a predator in aggregate. In addition, they "pulse" by moving their bodies up and down on their long legs, and this behavior could deter would-be attackers. Some aggregations can include hundreds of individuals!

Pseudoscorpions (Pseudoscorpionida)

Due to their small size (2 to 7 mm), pseudoscorpions (also called false scorpions or book scorpions) are rarely seen in the garden. Given their scorpion-like appearance, they may cause unnecessary alarm if found, even though these tiny predators pose no threat to humans. Pseudoscorpions are brown to tan, have four pairs of legs attached to their cephalothorax, and a segmented abdomen. Most species have two to four eyes, and all have characteristic enlarged pedipalps that resemble those of a scorpion. These pedipalps are used to capture and kill prey, disperse, and build nests. Like spiders, pseudoscorpions can spin silk, but they do not use it to capture prey. Silk is used to construct shelters, in mating, molting, egg laying, development, and as protection from cold temperatures.

Courtship varies between species; in some, males deposit multiple spermatophores in the presence or even absence of a female, and once the female locates it, she uses it to fertilize her eggs. In other species, courtship is more complex. Some pseudoscorpions perform a courtship dance where the male grasps the pedipalps of the female in his and guides her back and forth over a spermatophore to fertilize her eggs. Most females will carry their eggs as they develop in a brood pouch, and females may spend this time in a silken retreat. After the eggs hatch, the young ride on their mother for a short time. Pseudoscorpions develop through five immature stages before molting to an adult, sometimes taking over a year to complete their development.

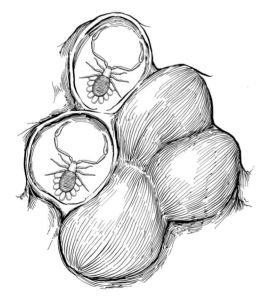

△ *Pseudoscorpions create igloo-shaped shelters from silk where they retreat for mating, molting, laying eggs, and to gain some protection from freezing temperatures.*

Photo: D. Shetlar

△ *Pseudoscorpions have four pairs of legs, a segmented abdomen, and enlarged pedipalps that look similar to scorpions. These small creatures may look frightening, but they do not bite or sting humans.*

△ *Pseudoscorpions use phoresy, or dispersal via other animals, to move around. This individual is grasping the leg of a small fly.*

Given their small size and lack of wings, dispersal might seem difficult for pseudoscorpions. But they've found a way around these challenges by becoming excellent hitchhikers! Pseudoscorpions use phoresy – dispersal via other animals – to move around. They can disperse by grasping onto a fly, beetle, moth, harvestmen, spider, or other arthropod. In some cases, the pseudoscorpions will even provide a service to their transporting arthropod by feeding on its parasites.

It is not surprising that pseudoscorpions feed on small prey. They commonly consume springtails (Collembola), flies (Diptera), thrips (Thripidae), spider mites (Tetranychidae), ants (Formicidae), and beetle larvae (Coleoptera). Most are active predators that forage on plant foliage and the ground surface for these prey, but some use a sit-and-wait strategy and grasp prey that inadvertently wanders within reach. Beyond the garden, pseudoscorpions can sometimes be found inside the house. The book scorpion, *Chelifer cancroides*, feeds on clothes moths (Tineidae: *Tineola*), carpet beetles (Dermestidae), and book lice (Psocoptera), household pests that feed on clothing and the glue used to bind books!

Scorpions (Scorpiones)

There are several families of scorpions that occur within the United States, primarily in the western states, but also in the southeast of the country, with a distribution as far north as Kentucky. Scorpions are not necessarily common in gardens, but in their geographic range, they can be found within residential landscapes and in nearby natural areas. Most scorpions are yellow to brown and range from ½ to 4¾ inches (1.5 to 12 cm) in body length. They have up to twelve eyes on their cephalothorax, with one particularly large central pair. They are easy to identify using two characteristic features. First, they have enlarged, clawlike pedipalps. The pincers at the end of the pedipalp can be long and slim or thick and crablike. Second, their abdomen narrows to a bulb-shaped segment called the telson, which contains venom glands, and ends with a curved spine or stinger that is used to inject the scorpion's venom.

All scorpions are venomous, but that does not mean that they are aggressive or dangerous. Worldwide, only a small number of scorpions are dangerous to humans, and only one U.S. species poses a threat, the Arizona bark scorpion (*Centruroides sculpturatus*) in the family Buthidae. It is uniformly tan with slender palps and a long, slender abdomen, which is held coiled to the side of the scorpion at rest. This species is common in the Sonoran Desert in the southwestern United States and northern Mexico. Ranging from 2½ to 3 inches (6.5 to 8 cm) in body length, the Arizona bark scorpion is active at night and spends the day under rocks, in the crevices of tree bark, or in wood piles. Others in this family include the striped bark scorpion (*Centruroides vittatus*) and the Florida bark scorpion (*Centruroides gracilis*). These species are considered less dangerous to humans. It is always a good idea to seek medical attention if stung by any scorpion, however. Reactions to their venom do vary, and making an identification after being stung is not typically one's top priority!

The family Caraboctonidae includes the giant hairy scorpions in the genus *Hadrurus*. The Arizona desert hairy scorpion (*Hadrurus arizonensis*) is the largest species in the United States at 4¾ inches (12 cm) in body length! Despite its large size, the venom of this species is not nearly as potent as that of the Arizona bark scorpion. Size is not a good predictor of venom potency in these predators.

Forest scorpions in the family Vaejovidae range from ¾ to 4 inches (2 to 10 cm) and are found throughout the Appalachian region and the western United States. This family includes many species, such as the California forest scorpion (*Uroctonus mordax*), giant sand scorpion (*Smeringurus mesaensis*), northern scorpion (*Paruroctonus boreus*), southern unstriped scorpion (*Vaejovis carolinianus*), California common scorpion (*Paruroctonus silvestrii*), and sawfinger scorpions in the genus *Serradigitus*.

Other families found in the southwestern states include the toothed scorpions (Diplocentridae) and one species in the family Superstitioniidae. The superstition mountains scorpion (*Superstitionia donensis*) is named for the area where it was first collected near Phoenix, Arizona.

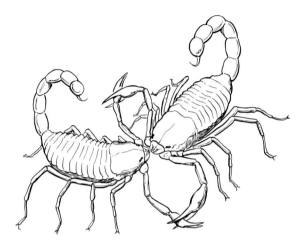

△ *Scorpions engaged in their courtship dance or "promenade á deux." The male grasps the female's pedipalp in his, sometimes piercing it to inject a small amount of venom—a precautionary measure to ensure that the female does not become aggressive and attack her mate. He may also "kiss" her by grasping her chelicera with his. They will move around together as he looks for a suitable place to mate. Once he finds the perfect location, the male will deposit a spermatophore, or capsule containing sperm, on the ground. He will then guide the female over it so that it is taken up into her genital opening, allowing fertilization to take place.*

Scorpions have a very complex courtship that involves a mating dance called the "promenade á deux," where the male grasps a female with his pedipalps and moves her around looking for a suitable place for mating. Males will deposit a spermatophore, or capsule containing sperm, on the ground and will guide the female over it to ensure that spermatophore is taken up into the female's genital opening, which triggers the release of sperm and allows fertilization to take place. During this dance, males

△ *The Arizona bark scorpion* Centruroides sculpturatus *is found in the southwestern United States.*

△ *The giant hairy scorpions* (Hadrurus) *include some of the largest species found in the United States. This male and female pair are engaged in a promenade á deux, or courtship dance.*

△ *Many species of forest scorpions* (Vaejovidae), *including the giant sand scorpion* (Smeringurus mesaensis), *are found in the United States.*

may pierce the female's pedipalp, possibly injecting a small amount of venom and may also "kiss" her by grasping her chelicera with his. It is thought that these behaviors reduce the likelihood that the female will become aggressive and attack her mate. After mating, the male retreats quickly to avoid being eaten. Pregnancy in scorpions can last two or more months, and females give birth to live young instead of eggs. The immature scorpions will climb aboard their mother, who will protect them from predation from other scorpions. The length of time immature scorpions stay with their mother varies between species, and they will molt several times before reaching the adult stage.

Scorpions are generalist predators of arthropods, including many that are considered garden pests. Most scorpions do their hunting at night and typically only reject prey that emit a defensive odor. Some are active hunters, whereas others use a sit-and-wait strategy, remaining motionless until prey comes within the reach of their clawlike pedipalps. They use their pedipalps to grasp their prey and hold it while they macerate it with their chelicerae and consume it. Scorpions are voracious predators but can also go for long periods without food—up to a year in some species!

Whip Scorpions (Thelyphonida)

Photo: D. Shetlar

△ *Only the giant vinegaroon* (Mastigoproctus giganteus) *is found in the United States. This interesting creature is active at night using its antenniform legs to detect odors or vibrations from unsuspecting prey.*

Whip scorpions, also called vinegaroons, look somewhat similar to whip spiders (Amblypygi), with the addition of a long "tail" called a flagellum, a sensory organ that aids in the detection of prey or potential predators. The name whip scorpion comes from their resemblance to scorpions—like the scorpion, this species uses their enlarged clawlike pedipalps with many spines to grasp and macerate their prey.

These interesting arthropods are found throughout the tropics and subtropics, but only the giant vinegaroon (*Mastigoproctus giganteus*) resides in the United States. The common name vinegaroon comes from their ability to emit a foul odor. If threatened, they will spread their pedipalps, lift their abdomen, and spray their attacker with acetic acid. This arthropod ranges in body length from 5 to 7 inches (12.5 to 17.5 cm) and can be found in Arizona, Florida, New Mexico, Oklahoma, and Texas. Giant vinegaroons are dark brown with eight to twelve eyes arranged in three groups on their prosoma, but they have poor vision. They rely on their first pair of thin antenniform legs to detect stimuli such as odors and vibrations from their surroundings. As the animal forages for prey, it taps these legs on the surface of the soil. It also looks for mates using its other three pairs of walking legs to move around in the environment.

Like many arachnids, whip scorpions have an interesting courtship dance that can last several hours. During this time, the male approaches the female and shakes his antenniform legs. Males will then grasp the female's pedipalp or thin first pair of legs with his pedipalps and tap his antenniform legs on her legs and pedipalps, as well as on the ground surface. As with many

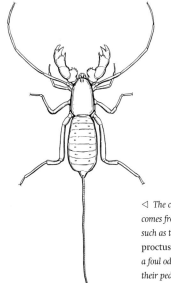

◁ *The common name "vinegaroon" comes from the ability of whip scorpions, such as the giant vinegaroon* Mastigoproctus giganteus, *to produce and emit a foul odor. If threatened, they will spread their pedipalps, lift their abdomen, and spray their attacker with acetic acid!*

arachnids, the male ends this dance by depositing a spermatophore on the ground to be taken up by the female. Females lay their eggs within a burrow they construct, and when immature whip scorpions hatch, they spend the first part of their lives on their mother's abdomen. Whip scorpions are active at night and will consume many different arthropods, including garden pests. They spend the day concealed in their burrow.

Camel Spiders (Solifugae)

The camel spider's most obvious trait is its very large jaws! These arachnids use their chelicerae in prey capture, burrow construction, defense, and mating. A number of urban legends have circulated about camel spiders, but most are exaggerations or completely false. Beginning during the Gulf War in the early 1990s, altered photographs of soldiers with camel spiders made them appear to be as long as a human arm. In some parts of the world, camel spiders can be quite large (up to 6 inches [15 cm]), but they are not nearly as large in the United States, where they grow up to 1 inch (2.5 cm) in length. Legend also has it that this spider can run as fast as 25 mph (40 kph). In reality, they are fast, moving at a rate of up to 10 mph (16 kph), but not as fast as some would have you believe. And, they certainly do *not* attack and kill humans as many urban legends claim! They might bite if handled, but camel spiders pose no serious risk.

Their common name does not come from the fact that they eat the stomachs of camels, either. Instead, it is due to an arched cephalothorax in some species that resembles a camel's hump. They have two large eyes at the center of their cephalothorax and a second pair of small eyes on its edge. Camel spiders have leglike pedipalps and four pairs of actual legs. But the first pair of legs is not used for walking; sensory structures located on these legs help detect odors of prey or potential predators. These spiders are more likely to rely on the detection of chemical and vibrational stimuli than vision. Because of how fast camel spiders can run, they are able to cover a significant area searching for prey and mates. They hunt on the ground but can also climb vegetation in search of prey.

Males actively search for mates, and when they encounter a female, they touch her with their pedipalps, causing her to enter a torporlike frozen state; this behavior may prevent the female from attacking the male. The male may pick the female up while she is in this state and move her a short distance before mating, which occurs when the male inserts a spermatophore into the female with his chelicerae. After this, he retreats quickly to avoid being eaten! The female either finds or constructs a burrow in which to lay her eggs. She will then leave her burrow and cover the entrance to protect the eggs from potential predators. These eggs hatch in four to six weeks and will molt up to ten times before reaching adulthood. Solifugae have only one generation per year.

△ *Camel spiders (Solifugae) have large chelicera and leglike pedipalps that make it appear as if they have five pairs of legs. These arachnids can run very fast and are most common in the desert southwest.*

In the United States, there are two families of Solifugae—the curve-faced camel spiders (Ammotrechidae) and the straight-faced camel spiders (Eremobatidae). Both families are most common in the desert southwest, where they are found in sandy open habitats such as deserts and dry grasslands. They are not a common sight in the garden, but you may encounter this interesting arthropod if your home is near these types of habitats. Most solifugids are generalist predators that consume a variety of different arthropods, only rejecting prey that is distasteful due to the presence of chemical defensive compounds. They are extremely voracious, and females have been known to consume prey to the point that their abdomen is distended, making movement difficult. Some species have a more specialized diet of termites, ants, or honey bees. The species *Eremobates pallipes* have been reported to feed on bed bugs, but the likelihood of a biological control program utilizing camel spiders inside homes to control this pest seems doubtful.

12

Centipedes (Chilopoda)

As a gardener, it's nearly impossible to avoid seeing these creepy crawly critters out in the yard. Turn over a rock, and you're likely to see one scurrying away—a centipede. These elongated, flattened, multi-segmented arthropods have one set of legs per body segment, and the number of body segments varies between species, ranging from 15 to 191. Do the math—that means centipedes can have between 30 and 382 legs. With all those legs, it's not surprising that centipedes are fast-moving predators. The first pair of legs, called forcipules, is modified into "poison claws," which act like pincers and have ducts near the tip of the claw that release venom. These ducts are unique to centipedes and are used to capture and immobilize their prey. Although centipedes have eyes, they are only able to detect light intensity. Instead of finding prey by sight, they hunt using their antennae, which detect odor cues from their prey. Centipedes feed on a diversity of insects, worms, and other soil invertebrates, but they are also known to feed on plant material. They can survive long periods without eating, sometimes up to six months.

Courtship Rituals and Reproduction

Like many of the arachnids discussed in chapters 10 and 11, centipedes also reproduce via indirect mating. In some centipede species, males will place spermatophores on a silk pad that he spins for females to find. In other species, males will initiate a courtship dance with his prospective mate, which can last for hours. He will tap her hind legs with his antennae to encourage her to move toward the silk pad and take up the spermatophore. Some centipedes are also able to reproduce parthenogenically, a process of self-fertilization that does not require mating between a male and female.

Females lay their eggs, either singularly or in groups, under rocks or logs, in the soil, or under bark. In some species, females guard the eggs and newly hatched young. Immature individuals look like miniature versions of adults and undergo several molts before they reach the adult stage. There are four centipede orders found within the United States.

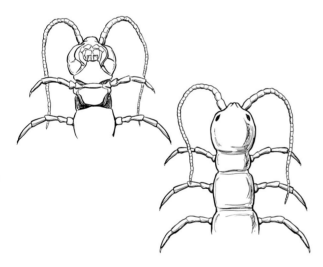

△ Centipedes are the only insects with the first pair of legs modified into pincers called forcipules or "poison claws." Centipedes use their poison claws to inject toxic venom into their prey and as a means of defense against their predators.

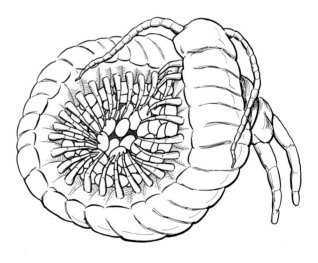

△ This female centipede is guarding her developing eggs. Females lay their eggs in protected areas, such as under rocks or in the soil.

Soil Centipedes (Geophilomorpha)

Soil centipedes constitute a large order with several families. These centipedes are elongated and eyeless with 27 to 191 pairs of legs and range in length from ½ to 8 inches (1.5 to 20 cm). Despite their many legs, this is the only order of centipedes that are not fast-moving. They are light tan to dark brown and are commonly found in rotting logs and within leaf litter and soil. In addition to consuming soil-dwelling pests, soil centipedes perform another beneficial service to gardeners; they burrow in the soil much like earthworms, providing aeration that allows water and nutrients to reach plant roots. If threatened, soil centipedes coil up and exude a defensive compound to ward off predators.

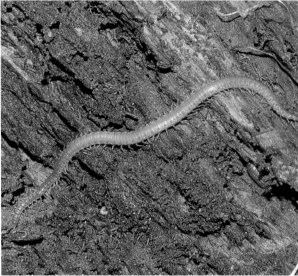

Photo: D. Shetlar

Stone Centipedes (Lithobiomorpha)

This order includes two families found in the United States, Henicopidae and Lithobiidae. Adults are ¾ to 2 inches (2 to 5 cm) in length, typically brown, and have fifteen pairs of legs and eighteen body segments. Stone centipedes are commonly found in gardens on tree bark, under stones or logs, and in leaf litter and mulch. Some species are known to climb trees, where they feed on aphids (Aphidiadae) and scales (Coccoidea). If threatened, these centipedes use their hind legs to flick a sticky material onto their attacker.

Photo: D. Shetlar

House Centipedes
(Scutigeromorpha)

House centipedes are the most distinctive-looking order of centipedes. Adults have fifteen pairs of very long legs, bodies as long as 1¼ inches (3 cm), and long antennae. In females, the last pair of legs is very long—nearly twice as long as the body. Their body is light in color with dark stripes running down the center and each side of its body. House centipedes have a rounded head and compound eyes. The species *Scutigera coleoptrata*, an exotic species that has been introduced into the United States, is commonly found inside houses. These centipedes forage for a number of pests, including silverfish (Lepismatidae), carpet beetles (Dermestidae), and cockroaches (Blattodea).

Photo: D. Shetlar

Although they are very creepy, house centipedes do not pose a risk to people and will flee quickly if threatened.

Tropical Centipedes
(Scolopendromorpha)

Tropical centipedes are the largest centipedes in the United States, with either twenty-one or twenty-three body segments and ranging in length from 1 to 6 inches (2.5 to 15 cm). This species is found mainly in the southern states, although there are a couple of species that can be found as far north as Canada. Tropical centipedes are relatively fast moving and can construct burrows to escape unfavorable temperatures. Species found within the United States include the giant redheaded centipede (*Scolopendra heros*) and Florida blue centipede (*Hemiscolopendra marginata*). One other characteristic to note: This order has a very painful bite.

Photo: D. Shetlar

References

Agosti, D., J. Majer, L. Alonso, & T. Schultz. (2000). *Ants: Standard Methods for Measuring and Monitoring Biodiversity.* Washington: Smithsonian Institution Press.

Aldrich, J. (1894). "Courtship Among the Flies." *The American Naturalist: Vol. 28,* 35-37.

American Arachnological Society. *About Arachnids.* (n.d.) Retrieved from: www.americanarachnology.org/education/about_arachnids.html.

Andrewes, C. (1969). *The Lives of Wasps and Bees.* London: Chatto & Windus.

Beccaloni, J. (2009). *Arachnids.* Berkeley: University of California Press.

Bessin, R. and J. Obrycki. (2011). "An IPM scouting guide for natural enemies of vegetable pests." University of Kentucky Extension. Bulletin ENTFACT-67.

Boone, J. (2014). *Invertebrates Around Las Vegas.* Retrieved from Bird and Hike: www.birdandhike.com/Wildlife/Invert/_Inverts.htm.

Borror, D., & White, R. (1970). *A Field Guide to Insects: American North of Mexico.* Boston: Houghton Mifflin Company.

Borror, D., C. Triplehorn, & N. Johnson. (1989). *An Introduction to the Study of Insects, Sixth Edition.* Fort Worth: Harcourt Brace College Publishers.

Bouchard, R.W., Jr. (2004). "Guide to aquatic macroinvertebrates of the Upper Midwest." Water Resources Center, University of Minnesota, St. Paul, MN.

Brookhart, I., & Brookhart, J. (2006). "An Annotated Checklist of Continental North American Solifugae with Type Depositories, Abundance, and Notes on Their Zoogeography." *The Journal of Arachnology: Volume 34,* 299-329.

Capinera, J. (2004). *Encyclopedia of Entomology (Vols. I, II, III).* Dordrecht, Netherlands: Kluwer Academic Publishers.

Coll, M., & J. Ruberson. (1998). *Predatory Heteroptera: Their Ecology and Use in Biological Control.* Lanham: Entomological Society of America.

Cornell University: College of Agriculture and Life Sciences. (n.d.). *Biological Control: A Guide to Natural Enemies in North America.* Retrieved from www.biocontrol.entomology.cornell.edu/index.php.

Cranshaw, W. (2004). *Garden Insects of North America: The Ultimate Guide to Backyard Bugs.* Princeton: Princeton University Press.

Debach, P., & D. Rosen. (1991). *Biological Control by Natural Enemies.* Cambridge, UK: Cambridge University Press.

Den Boer, P., M. Luff, D. Mossakowski, & F. Weber. (1986). *Carabid Beetles: Their Adaptations and Dynamics .* Stuttgart, Germany: Gustav Fischer.

Dolling, W. (1991). *The Hemiptera.* Oxford, UK: Oxford University Press.

Erwin, T., G. Ball, D. Whitehead, & A. Halpern. (1979). *Carabid Beetles: Their Evolution, Natural History, and Classification.* The Hague, Netherlands: Dr. W. Junk Publishers.

Evans, A., & Bellamy, C. (1996). *An Inordinate Fondness for Beetles.* New York: Henry Holt and Company.

Evans, G. (1975). *The Life of Beetles.* London, UK: George, Allen & Unwin Ltd.

Flint, M. & S. Driestadt. (1999). *Natural Enemies Handbook: The Illustrated Guide to Biological Pest Control.* Oakland: University of California Press.

Fuester, R., R. Casagrande, R. Van Driesche, M. Mayer, D. Gilrein, L. Tewksbury & H. Faubert. (2009). *Birch Leafminer: A Success.* Retrieved from Northern Research Station: www.nrs.fs.fed.us/pubs/gtr/gtr-nrs-p-51papers/13fuester-p-51.pdf.

Gauld, I. & B. Bolton (1988). *The Hymenoptera.* Oxford, UK: Oxford University Press.

Gibbons, J. (1979). "A Model for Sympatric Speciation in Megarhyssa (Hymenoptera: Ichneumonidae): Competitive Speciation." *The American Naturalist: Vol. 114,* 719-741.

Hajek, A. (2004). *Natural Enemies: An Introduction to Biological Control.* Cambridge, UK: Cambridge University Press.

Harvey, M. (2003). *Catalogue of the Smaller Arachnid Orders of the World.* Collingwood, Australia: CSIRO Publishing.

Henry, T. & R. Froescnher. (1988). *Catalog of the Heteroptera, or True Bugs, of Canada and the Continental United States.* Leiden, Netherlands: E.J. Brill.

Hillyard, P. & J. Sankey. (1989). *Harvestman: Keys and Notes for Identification of the Species.* Leiden, Netherlands: E.J Brill.

Hodek, I. (1973). *Biology of Coccinellidae.* The Hague, Netherlands: Klewer Academic Publishers.

Holland, J. (2002). *The Agroecology of Carabid Beetles.* Maryland: Intercept.

Iowa State University: Department of Entomology. (n.d.). *Bug Guide.* Retrieved from www.bugguide.net.

Johnson, J.B. (2000). "Adult Insect Taxonomy Lecture Notes." University of Idaho, Department of Plant, Soil, and Entomological Sciences.

Lindroth, C. H. (1969). *The Ground-Beetles of Canada and Alaska.* Lund, Sweden: Berlingska Boktryckeriet.

Lundren, J. (2009). *Relationships of Natural Enemies and Non-Prey Foods. Progress in Biological Control Vol. 7.* Dordrecht, Netherlands: Springer.

Lundren, J., J. Duan, M. Paradise & R. Wiedenmann. (2005). "Rearing Protocol and Life History Traits for *Poecilus chalcites* (Coleoptera: Carabidae) in the Laboratory." *Journal of Entomological Science: Vol. 40,* 126-135.

Mader, E., M. Shepherd, V. Mace, S. Black & G. LeBuhn. (2011). "Attracting Native Pollinators: Protecting North America's Bees and Butterflies." *The Xerces Society Guide.* North Adams: Storey Publishing.

Majerus, M. & P. Kearns. (1989). *Ladybirds.* Slough, UK: Richmond Publishing Co. Ltd.

Marshall, S. (2006). *Insects: Their Natural History and Diversity: with a Photographic Guide to Insects of Eastern North America.* Buffalo: Firefly Books.

Michigan State University Extension. (n.d.). *Native Plants and Ecosystem Services.* Retrieved from www.nativeplants.msu.edu.

Minelli, A. (1978). "Secretions of Centipedes." *Arthropod Venoms.* (S. Bettini ed.). 611 pgs.

Missouri Department of Conservation. (n.d.) *Field Guide.* Retrieved from mdc.mo.gov/discover-nature/field-guide.

New, T. (2010). *Beetles in Conservation.* West Sussex, UK: Wiley-Blackwell.

Noonan, G. (1991). *Classification, Cladistics, and Natural History of Native American Harpalus Latrielle (Insecta: Coleoptera: Carabidae: Harpalini) Excluding Subgenera Glanodes and Psudeophonus.* Lanham: Entomological Society of America.

O'Neill, K. (2001). *Solitary Wasps: Behavior and Natural History.* Ithaca: Comstock Publishing Associates.

Papp, C. (1984). *Introduction to North American Beetles.* Sacramento: Entomography Publications.

Pearson, D., C. Knisley & C. Kazilek. (2006). *A Field Guide to the Tiger Beetles of the United States and Canada.* Oxford, UK: Oxford University Press.

Peckham, G.W. and E.G. Peckham. (1905). *Wasps Social and Solitary.* Boston and New York: Houghton, Mifflin & Company.

Penn State University: Department of Entomology. (n.d.). *Entomology Fact Sheets.* Retrieved from Penn State University: www.ento.psu.edu/extension/factsheets.

Pickett, C., & Bugg, R. (1998). *Enhancing Biological Control: Habitat Management to Promote Natural Enemies of Agricultural Pests.* Oakland: University of California Press.

Punzo, F. (1998). *The Biology of Camel-Spiders (Arachnida, Solifugae).* Boston: Kluwer Academic Publishers.

Rein, J. (n.d.). *The Scorpion Files.* Retrieved from www.ntnu.no/ub/scorpion-files/index/php.

Resh, V. & R. Carde. (2003). *Encyclopedia of Insects.* Amsterdam, Netherlands: Academic Press.

Rico-Gray, V. & P. Oliveira. (2007). *The Ecology and Evolution of Ant-Plant Interactions.* Chicago: The University of Chicago.

Schuster, R. & P. Murphy. (1991). *The Acari: Reproduction, Development, and Life History Strategies.* London, UK: Chapman & Hall.

Shelley, R. (2002). *A Synopsis of the North American Centipedes of the Order Scolopendromorpha (Chilopoda).* Martinsville: Virginia Museum of Natural History Memoir.

Shelley, R. (n.d.). *The Myriapods, The Worlds Leggiest Animals.* Retrieved from The Myriapoda (Millipedes, Centipedes) Featuring the North American Fauna: www.nadiplochilo.com.

Spoczynska, J. (1975). *The World of Wasp.* New York: Crane, Russak & Company.

Spradbery, J. (1973). *Wasps.* Seattle: University of Washington Press.

Stork, N. (1990). *The Role of Ground Beetles in Ecological and Enviromental Studies.* Andover: Intercept Ltd.

Sudd, J., & N. Franks. (1987). *The Behavioural Ecology of Ants.* Glasgow: Blackie & Son Ltd.

Symondson, W., D. Glen, M. Erickson, J. Lidell & C. Langdon. (2000). "Do Earthworms Help to Sustain the Slug Predator *Pterostichus melanarius* (Coleptera: Carabidae) Within Crops? Investigations Using Monoclonal Antibodies." *Molecular Ecology: Vol. 9,* 1279–1292.

Tappey, H.J., W.E. Conner, J. Meinwald, H.E. Eisner & T. Eisner. (1976). "Benzoyl cyanide and mandelonitrile in the cyanogenetic secretion of a centipede." *Journal of Chemical Ecology 4:* 421-429.

Thiele, H. (1977). *Carabid Beetles in Their Enviroment.* Berlin, Germany: Springer-Verlag.

University of California. (n.d.). *Agriculture and Natural Resources.* Retrieved from UC IPM Online: www.ipm.ucdavis.edu.

University of Florida. (n.d.). *Beneficial Organisms (IN).* Retrieved from IFAS Extension: edis.ifas.ufl.edu/topic_in_beneficial_organisms.

University of Kentucky: Department of Entomology. (2010). *Kentucky Critter Flies.* Retrieved from University of Kentucky: www.uky.edu/Ag/CritterFiles/casefile/casefile.htm.

University of Minnesota. (n.d.). *Insect Factsheets.* Retrieved from University of Minnesota Extension: www.extension.umn.edu/garden/insects.

University of Minnesota. (n.d.). *VegEdge.* Retrieved from University of Minnesota Extension: www.vegedge.umn.edu.

University of Wisconsin-Madison: Department of Entomology. (n.d.). *Biological Control News.* Retrieved from University of Wisconsin-Madison:Department of Entomology: www.entomology.wisc.edu/mbcn/index.html.

Van Driesche, R., & Bellows, T. (1996). *Bilogical Control.* New York: Chapman & Hall.

Virginia Tech University. (n.d.). *The Mid-Atlantic Regional Fruit Loop: The Virginia Fruit Page.* Retrieved from Virginia Tech: www.virginiafruit.ento.vt.edu.

Xerces Society for Invertebrate Conservation. (n.d.) *Bee-Friendly Plant Lists.* Retrieved from: www.xerces.org.

Williams, S., & Hefner, R. (1918). "The Millipedes and Centipedes of Ohio." *Ohio Biological Survey Vol. 4,* 93-146.

Williston, S.W. (1896). *Manual of the Families and Genera of North American Diptera.* New Haven: Hathaway.

Wygoldt, P. (2000). *Whip Spiders (Chelicerata: Amblypygi): Their Biology, Morphology and Systematics.* Stenstrup, Denmark: Apollo Books.

Index

About the Author

Mary M. Gardiner is an associate professor in the department of entomology at The Ohio State University. She received a B.S. in resource ecology and management from the University of Michigan, an M.S. in entomology from the University of Idaho, and a Ph.D. in entomology from Michigan State University.

Mary is originally from small community in northern Michigan, where she grew up gardening with her family and enjoying the outdoors. Her love of nature developed into a career studying how human activity influences the sustainability and ecological function of agricultural and urban habitats. Much of her current research takes place within Cleveland, Ohio, a city that encompasses 20,000 vacant lots where homes and businesses once stood. Here, her laboratory is studying how the redesign of vacant land to restore native plant communities, improve storm water infiltration, and provide access to locally-produced food influences the environmental quality of city neighborhoods. Mary is also a state extension specialist who is active in the Ohio Master Gardener Program. She regularly presents programs on arthropod identification and how to enhance home landscapes, urban green spaces, and small-scale farms as habitats for beneficial arthropods.

Acknowledgments

I have been fortunate to have many outstanding mentors during my academic career. In particular I want to thank John Witter for inspiring me to go into the field of entomology and Jim Barbour and Doug Landis for teaching me how to be a scientist and supporting me though my graduate career. I would also like to acknowledge James "Ding" Johnson for his excellent adult insect identification course at the University of Idaho. I referred to my notes from this class frequently.

I want to acknowledge all of my students and colleagues at The Ohio State University and across the country for all of their support, feedback, and ideas. Thank you to James Harwood, University of Kentucky, for his review of the Spiders and Other Arachnids chapters, and graduate students Erin O'Brien and Andrea Kautz, The Ohio State University, for their reviews of the True Bugs and Predator and Parasitoid Flies chapters. Thank you to graduate student Ray Fisher, University of Arkansas, for his contributed text and a review of the Acari section within the Other Arachnids chapter. Undergraduate Clint Fleshman collected and organized all the photographs and graduate student MaLisa Spring read each chapter for errors, including double checking all the scientific names.

Writing a book about predator and parasitoid arthropods that covered important taxa across the United States was a daunting task. I felt it was important to include both common species as well as some less common arthropods that held a certain level of fascination for the gardener. I had lots of ideas, but the website BugGuide.net served as a great tool in this process. This site, which is managed by Iowa State University, allows users to upload arthropod photographs which are identified by experts. The number of submitted images was one very helpful criterion I used to select the arthropods to discuss.

The majority of the photos in this book were taken by photographers who participate on BugGuide.net. These individuals range from professional photographers to researchers and home gardeners who saw an interesting insect and captured an amazing shot. In addition to photographers I contacted through BugGuide.net, I received many of the images in the Spider and Other Arachnids chapters from graduate student Sarah Jane Rose, The Ohio State University. My colleague Dave Shetlar, better known as The Bug Doc, and graduate students MaLisa Spring and Ray Fisher, also provided several great photos. Clearly this book could not have been written without the contributions of many outstanding photographers, and I want to thank each of them.

This is my first book, and I certainly learned a lot about the publication process. I want to thank editor Kari Cornell, creative director Regina Grenier, project manager Renae Haines, and editorial director Mary Ann Hall for all of their guidance and feedback throughout this project. It was also a pleasure to work with Michael Cooley, the artist and instructor who created the illustrations.

Finally, a big thank you to my family for always supporting my love of insects and gardening!

635.0496 Gardiner, Mary M.
G
          Good garden bugs.